To Brenda and Hazel

Beginning Logic

Beginning Logic

E.J. Lemmon

CHAPMAN & HALL/CRC

Boca Raton   London   New York   Washington, D.C.

Library of Congress Cataloging-in-Publication Data

Catalog record is available from the Library of Congress.

First edition 1965
Reprinted 1967, 1968, 1971, 1972, 1977 (twice), 1979, 1983, 1984, 1986 (twice)
Second edition 1987
Reprinted 1990, 1991 published by Chapman & Hall Ltd, 1993
Reprinted 1994, 1997
First CRC Press reprint 1998
Originally published by Chapman & Hall

No claim to original U.S. Government works
International Standard Book Number 0-412-38090-0
Printed in the United States of America   1   2   3   4   5   6   7   8   9   0
Printed on acid-free paper

CONTENTS

Contents

CHAPTER 4   THE PREDICATE CALCULUS 2

PREFACE

The material for this book has grown out of lecture-courses given in the University of Oxford between 1958 and 1962 and in the University of Texas during the spring semester of 1961, where indeed I chiefly found the time to write it. It has also subsequently been used for a one-semester course in Claremont, California, and for a summer-school course at the University of California, Los Angeles. The book is intended in the first place for use in connection with one-term or one-semester introductory logic courses at university level; but I see no reason why it could not be employed in schools, and it is my hope and expectation that logic will increasingly be taught at school level. No prior knowledge of either philosophy or mathematics (except the ability to count and to recognize some elementary algebraic equations) is assumed, and the book is addressed to those who find mathematical thinking difficult rather than to those who find it easy. Thus an intelligent layman working on his own should find it in large part comprehensible.

TO THE STUDENT

The aim of the book is to provide the student with a good working knowledge of the propositional and predicate calculi—the foundations upon which modern symbolic logic is built. Accordingly, emphasis is placed on the actual technique of proof-discovery. Within this framework, I have tried to sacrifice formal accuracy as little as possible to intuitive plausibility, though on occasion, and in the early stages, this has sometimes quite deliberately been done. The result is probably dry reading but should not be hard going. On the other hand, I have included certain more theoretical sections, partly to indicate the style of more advanced logic and partly as a bait for the better student who might be inclined to pursue the subject further (the bibliography and notes following it are designed to guide him in this regard). In particular, Chapter 2, Sections 4 and 5, and

Chapter 4, Section 2, are a good deal more difficult than the rest of the book, and should be intelligently skipped (i.e. read through quickly) by the ordinary reader: nothing in what follows them hinges essentially on their contents.

In outline, we study in Chapter 1 some elementary proofs in the propositional calculus, and acquire familiarity with its rules of derivation. In Chapter 2, after mastering the vocabulary and grammar of this calculus, the student is introduced to truth-tables, which are then used as an independent control on the soundness and completeness of these rules. Chapter 3 presents the predicate calculus rules, and its basic results, in the same relatively informal way as the propositional calculus was treated in Chapter 1. Chapter 4 begins by *sketching* the theory of the predicate calculus (Sections 1 and 2), but continues with *applications* of it: first, with respect to identity; second, with respect to the traditional theory of the syllogism; third, with respect to properties of relations. Normal forms, which are part of the syllabus of many logic courses but which tend to receive scant attention in logic texts, are relegated to Appendix A (which can be read after Chapter 2, Section 3). Appendix B introduces the theory of classes, and may form a bridge between this and more advanced texts.

TO THE TEACHER

Natural deduction techniques are used throughout, and no mention is made of axiomatic developments of either calculus, though references are given in the bibliography. The manner of presentation of proofs owes most to Suppes [24],[1] who is followed in this respect by Mates [14]. The device of listing assumptions by number on the left of each line of a proof seems to me much clearer than more traditional approaches. The propositional calculus rules stem in essentials from Gentzen. They have the great merit that nearly all standard results can be obtained from them in proofs of not more than fifteen lines, whilst also being reasonably plausible to philosophically suspicious students. Thus they lend themselves to the generation of good exercises, but also keep the student within the confines of a clearly defined set of rules. Experts will notice that, if one half of the rule of double negation (' from— —A to derive A ') is

[1] Square bracket-references here and in the body of the text are to titles in the bibliography.

viii

dropped, the resultant set defines Johannson's minimal calculus, and, if to this set is added the rule ' from a contradiction to derive anything ', the resultant set defines the intuitionist propositional calculus. Good students become interested in these facts, once the implausibility of the law of excluded middle is suggested to them.

The predicate calculus rules again come from Gentzen, and are to be found in Fitch [4]; they reappear also in the recent book by Mates [14]. They have the property that, if added to the weaker sets of rules just mentioned, they yield the appropriate corresponding predicate calculus. A feature of our treatment that deserves mention is that the role normally played by free variables is here played by a different style of symbol, called an arbitrary name. Thus formation-rules become more complex than is usual; but such oddities as ' vacuous quantifiers ' disappear, and it proves possible to state quantifier-rules in a less restriction-infested form. This feature is not new: it goes back at least to Hilbert and Bernays, has been used by Hintikka, and appears in a variant form in Mates [14]. My experience has been that it causes students much less trouble than the more familiar notation.[1]

Any student worth his salt is going to be suspicious of the paradoxes of material implication. This fact counts strongly against *beginning* the treatment of the propositional calculus with the truth-table method. Accordingly, I have tried to woo the student in Chapter 1 into acceptance of a set of rules from which the paradoxes flow as natural consequences in Chapter 2; the truth-table method is then partly justified by appeal to these rules. Any teacher, therefore, who thinks that the paradoxes present *real* problems will (rightly) find my tactics underhand.

INTERNAL REFERENCES

Propositional calculus results are numbered 1–55, as they appear in Chapters 1 and 2. Predicate calculus results are numbered 100–165, also as they appear (Chapters 3 and 4). Certain results in the theory of classes are numbered 200–231 (Appendix B). Where such results are later used or referred to, I mention them by their number. Occasionally, the results in exercises are also mentioned; thus ' 2.4.1(*c*) ' refers to Exercise 1(*c*) of Chapter 2, Section 4. A number

[1] I should add that footnote 460 of Church [2] is critical of this device; but his case says nothing against its pedagogic attractiveness.

in brackets refers either to a line of a proof or to a sentence or formula so numbered earlier in the same section; context will always determine which.

＊

My thanks are due to Father Ivo Thomas, O.P., who read Chapter 1, and to Professor James Thomson, who read the whole book, for helpful comments and the correction of many errors. I am greatly indebted to my wife and to Miss Susan Liddiard for typing assistance, and to Mr Bruce Marshall for help with proof-reading and indexing. I owe a lot to discussions with colleagues about the best way to formulate logical rules: in particular, to Professor Patrick Suppes and Michael Dummett, whose idea it was (in 1957) that I should write this book. But my greatest debt is to the many students, in Oxford, Texas, and elsewhere, who forced me by their questions and complaints to write more clearly about the matters involved. The many faults of exposition that remain, of course, are mine.

I should like to dedicate this book to Arthur Prior, without whose encouragement and enthusiasm I would never have entered logic, and to the memory of my father, who I hope would have enjoyed it.

E. J. L.

Claremont, California
March 1965

CHAPTER 1

The Propositional Calculus 1

1 THE NATURE OF LOGIC

It is not easy, and perhaps not even useful, to explain briefly what logic is. Like most subjects, it comprises many different kinds of problem and has no exact boundaries; at one end, it shades off into mathematics, at another, into philosophy. The best way to find out what logic is is to do some. None the less, a few very general remarks about the subject may help to set the stage for the rest of this book.

Logic's main concern is with the soundness and unsoundness of arguments, and it attempts to make as precise as possible the conditions under which an argument—from whatever field of study—is acceptable. But this statement needs some elucidation: we need to say, first, what an argument is; second, what we understand by soundness; third, how we can make precise the conditions for sound argumentation; and fourth, how these conditions can be independent of the field from which the argument is drawn. Let us take these points in turn.

Typically, an argument consists of certain statements or propositions, called its *premisses*, from which a certain other statement or proposition, called its *conclusion*, is claimed to *follow*. We mark, in English, the claim that the conclusion follows from the premisses by using such words as ' so ' and ' therefore ' between premisses and conclusion. Instead of saying that conclusions do or do not follow from premisses, logicians sometimes say that premisses do or do not *entail* conclusions. When an argument is used seriously by someone (and not, for example, just cited as an illustration), that person is asserting the premisses to be true and also asserting the conclusion to be true *on the strength of* the premisses. This is what we mean by *drawing* that conclusion from those premisses.

Logicians are concerned with whether a conclusion does or does not follow from the given premisses. If it does, then the argument in question is said to be *sound*; otherwise *unsound*. Often the terms ' valid ' and ' invalid ' are used in place of ' sound ' and ' unsound '. The question of the soundness or unsoundness of arguments must be carefully distinguished from the question of the truth or falsity of the propositions, whether premisses or conclusion, in the argument. For example, a true conclusion can be soundly drawn from false premisses, or a mixture of true and false premisses: thus in the argument

> (1) Napoleon was German; all Germans are Europeans; therefore Napoleon was European

we find a true conclusion soundly drawn from premisses the first of which is false and the second true. Again, a false conclusion can be soundly drawn from false premisses, or a mixture of true and false premisses: thus in the argument

> (2) Napoleon was German; all Germans are Asiatics; therefore Napoleon was Asiatic

a false conclusion is soundly drawn from two false premisses. On the other hand, an argument is not necessarily sound just because premisses and conclusion are true; thus in the argument

> (3) Napoleon was French; all Frenchman are Europeans; therefore Hitler was Austrian

all the propositions are true, but no one would say that the conclusion followed from the premisses.

The basic connection between the soundness or unsoundness of an argument and the truth or falsity of the constituent propositions is the following: an argument cannot be sound if its premisses are *all* true and its conclusion false. A necessary condition of sound reasoning is that from truths only truths follow. This condition is of course not sufficient for soundness, as we see from (3), where we have true premisses and a true conclusion but not a sound argument. But, for an argument to be sound, it must *at least* be the case that if all the premisses are true then so is the conclusion. Now the logician is primarily interested in conditions for soundness rather than the actual truth or falsity of premisses and conclusion; but he may be secondarily interested in truth and falsity because of this connection between them and soundness.

What techniques does the logician use to make precise the conditions for sound argumentation? The bulk of this book is in a way a detailed answer to this question; but for the moment we may say that his most useful device is the adoption of a special symbolism, a logical notation, for the use of which exact rules can be given. Because of this feature the subject is sometimes called *symbolic* logic. (It is sometimes also called *mathematical* logic, partly because the rigour achieved is similar to that already belonging to mathematics, and partly because contemporary logicians have been especially interested in arguments drawn from the field of mathematics.) In order to understand the importance of symbolism in logic, we should remind ourselves of the analogous importance of special mathematical symbols.

Consider the following elementary algebraic equation:

$$(4) \quad x^2 - y^2 = (x + y)(x - y),$$

and imagine how difficult it would be to express this proposition in ordinary English, without the use of variables ' x ', ' y ', brackets, and the minus and plus signs. Perhaps the best we could achieve would be:

> (5) The result of subtracting the square of one number from the square of a second gives the same number as is obtained by adding the two numbers, subtracting the first from the second, and then multiplying the results of these two calculations.

Comparing (4) with (5), we see that (4) has at least three advantages over (5) as an expression for the same proposition. It is briefer. It is clearer—at least once the mathematical symbols are understood. And it is more exact. The same advantages—brevity, clarity, and exactness—are obtained for logic by the use of special logical symbols.

Equation (4) holds true for *any* pair of numbers x and y. Hence, if we choose x to be 15 and y to be 7, we have, as a consequence of (4):

$$(6) \quad 15^2 - 7^2 = (15 + 7)(15 - 7).$$

If we now compare (6) with (4), we can see that (6) is obtained from (4) simply by putting ' 15 ' in place of ' x ' and ' 7 ' in place of ' y '. In this way we can check that (6) does indeed follow from (4), simply by a glance to see that we have made the right substitutions for the variables. But if (6) had been expressed in ordinary English, as (4) was in (5), it would have been far harder to see whether it was soundly concluded from (5). Mathematical symbols make both the doing and the checking of mathematical calculations far easier. Similarly, logical symbols are humanly indispensable if we are to argue correctly and check the soundness of arguments efficiently.

If in the sequel it seems irritating that a special notation for logical work has to be learned, the reader should remember that he is only mastering for argumentation what he masters for calculation when he learns the correct use of ' $+$ ', ' $-$ ', and so on. This device, which logic has copied from mathematics, is the logician's most powerful tool for checking the soundness and unsoundness of arguments.

Our final question in this section is how the conditions for valid argument can be studied independently of the fields from which arguments are drawn: if this could not be done there would be no separate study called logic. A simple example will suffice for the moment. If we compare the two arguments

> (7) Tweety is a robin; no robins are migrants; therefore Tweety is not a migrant

and

> (8) Oxygen is an element; no elements are molecular; therefore oxygen is not molecular,

both of which are sound (one drawn from ornithology, the other from chemistry), it is hard to escape the feeling that they have something in common. This something is called by logicians their *logical form*, and we shall have more to say about it later. For the moment, let us try to analyse out in a preliminary way this common form. The first premiss of both (7) and (8) affirms that a certain particular thing, call it m (Tweety in (7), oxygen in (8)), has a certain property, call it F (being a robin in (7), being an element in (8)). The second premiss of (7) and (8) affirms that nothing with this property F has a certain other property, call it G (being a migrant in (7), being molecular in (8)). And the conclusion of (7) and (8) affirms that therefore the object *m* does not have the property *G*. We may state the common pattern of (7) and (8) as follows:

    (9) *m* has *F*; nothing with *F* has *G*;
         therefore *m* does not have *G*.

Once the common logical form has been extracted as in (9), a new feature of it comes to light. *Whatever* object *m* is picked out, *whatever* properties *F* and *G* are chosen to be, the pattern (9) will be valid: (9) as it stands is a pattern of a valid argument. For example, take *m* to be Jenkins, *F* and *G* to be the properties respectively of being a bachelor and being married: then (9) becomes

    (10) Jenkins is a bachelor; no bachelors are married;
         therefore Jenkins is not married,

which, like (7) and (8), is a sound argument. Yet (9) is not tied to any particular subject-matter, whether it be ornithology, chemistry, or the law; the *special* terminology—'migrant', 'molecular', 'bachelor'—has disappeared in favour of schematic letters '*F*', '*G*', '*m*'.

Form can thus be studied independently of subject-matter, and it is mainly in virtue of their form, as it turns out, rather than their subject-matter that arguments are valid or invalid. Hence it is the forms of argument, rather than actual arguments themselves, that logic investigates.

To sum up the contents of this section, we may define logic as the study, by symbolic means, of the exact conditions under which patterns of argument are valid or invalid: it being understood that validity and invalidity are to be carefully distinguished from the related notions of truth and falsity. But this account is provisional in the sense that it will be better understood in the light of what is to follow.

2 CONDITIONALS AND NEGATION

When we analyse the logical form of arguments (as we did in the last section to obtain (9) from (7) and (8)), words which relate to specific subject-matters disappear but other words remain; this residual vocabulary constitutes the words in which the logician is primarily interested, for it is on their properties that validity hinges.

Of particular importance in this vocabulary are the words 'if . . . then . . .', '. . . and . . .', 'either . . . or . . .', and 'not'. This chapter and the next are in fact devoted to a systematic study of the exact rules for their proper deployment in arguing. We have no single grammatical term for these words in ordinary speech, but in logic they may be called *sentence-forming operators on sentences*. I shall try to explain why they merit this formidable title.

In arguments, as we have already seen, propositions occur; an argument is a certain complex of propositions, among which we may distinguish premisses and conclusion. Propositions are expressed, in natural languages, in *sentences*. However, not all sentences express propositions; some are used to ask questions (such as 'Where is Jack?'), others to give orders (such as 'Open the door'). Where it is desirable to distinguish between sentences expressing propositions and other kinds of sentence, logicians sometimes call the former *declarative* sentences. Always, when I speak of sentences, it is declarative sentences I have in mind, unless there is some explicit denial. Now if we select two English sentences, say 'it is raining' and 'it is snowing', then we may suitably place 'if . . . then . . .', '. . . and . . .', and 'either . . . or . . .' to obtain the new English sentences: '*if* it is raining, *then* it is snowing', 'it is raining *and* it is snowing', and '*either* it is raining *or* it is snowing'. The two original sentences have merely been substituted for the two blanks in 'if . . . then . . .', '. . . and . . .', and 'either . . . or . . .'. Further, if we select one English sentence, say 'it is raining', then we may suitably place 'not' to obtain the new English sentence: 'it is *not* raining'. Thus, grammatically speaking, the effect of these words is to form new sentences out of (one or two) given sentences. Hence I call them sentence-forming operators on sentences. Other examples are: 'although . . . nevertheless . . .' (requiring two sentences to complete it), 'because . . . , . . .' (also requiring two), and 'it is said that . . .' (requiring only one).

(This book is written in English, and so mentions *English* sentences and words; but the above account could be applied, by appropriate translation, to all languages I know of. There is nothing parochial about logic, despite this appearance to the contrary.)

In this section we are concerned with the rules for manipulating 'if . . . then . . .' and 'not', and we begin by introducing special logical symbols for these operators. Suppose that P and Q are any two propositions; then we shall write the proposition that *if P then Q* as:

$$P \rightarrow Q.$$

Again, let P be any proposition; then we shall write the proposition that it is *not* the case that P as:

$$-P.$$

The proposition $P \rightarrow Q$ will be called a *conditional* proposition, or simply a *conditional*, with the proposition P as its *antecedent* and the proposition Q as its *consequent*. For example, the antecedent of the proposition that if it is raining then it is snowing is the proposition that it is raining, and its consequent is the proposition that it is snowing. The proposition $-P$ will be called the *negation* of P. For example, the proposition that it is not snowing is the negation of the proposition that it is snowing.

The letters 'P', 'Q', used here, should be compared with the variables 'x', 'y' of algebra; they may be considered as a kind of variable, and are frequently called by logicians *propositional variables*. In introducing the minus sign '$-$', I might say: let x and y be any two numbers; then I shall write the result of subtracting y from x as $x - y$. In an analogous way I introduced '$\rightarrow$' above, using propositional variables in place of numerical variables, since in logic we are concerned with propositions not numbers.

Propositional variables will also help us to express the logical form of complex propositions (compare the use of schematic letters 'F' and 'G' in (9) of Section 1). Consider, for example, the complex proposition

(1) If it is raining, then it is not the case that if it is not snowing it is not raining.

Let us use 'P' for the proposition that it is raining and 'Q' for the proposition that it is snowing. Then, with the aid of '$\rightarrow$' and '$-$', we may write (1) symbolically as:

(2) $P \rightarrow -(-Q \rightarrow -P)$

(we introduce brackets here in an entirely obvious way). (2), as well as being a kind of shorthand for (1), with the advantages of brevity and clarity—once at least the feeling of strangeness associated with novel symbolism has worn off—succeeds in expressing the logical

7

form of (1). We can see that (2) also gives the logical form of the quite different proposition

> (3) If there is a fire, then it is not the case that if there is not smoke there is not a fire:

here *P* is a stand-in for the proposition that there is a fire, and *Q* for the proposition that there is smoke.

When we argue, we draw or *deduce* or *derive* a conclusion from given premisses; in logic we formulate rules, called *rules of derivation*, whose object is so to control the activity of deduction as to ensure that the conclusion reached is validly reached. Another feature of ordinary argumentation is that it proceeds *in stages*: the conclusion of one step is used as a premiss for a new step, and so on until a final conclusion is reached. It will be helpful, therefore, if we distinguish at once between *assumptions* and *premisses*. By an *assumption*, we shall understand a proposition which is, in a given stretch of argumentation, the conclusion of *no* step of reasoning, but which is rather taken for granted at the outset of the total argument. By a *premiss*, we shall understand a proposition which is used, at a particular stage in the total argument, to obtain a certain conclusion. An assumption may be—and characteristically will be—*used* as a premiss at a given stage in an argument in order to obtain a certain conclusion. This conclusion may itself then be used as a premiss for a further step in the argument, and so on. Thus a premiss at a certain stage will be *either* an assumption of the argument as a whole *or* a conclusion of an earlier phase in the argument. At any given stage in the total argument, we shall have a conclusion obtained ultimately from a certain assumption or combination of assumptions, and we shall say that this conclusion *rests on* or *depends on* that assumption (those assumptions).

Roughly, our procedure in setting out arguments will be as follows. Each step will be marked by a new line, and each line will be numbered consecutively. On each line will appear *either* an assumption of the argument as a whole *or* a conclusion drawn from propositions at earlier lines and based on these propositions as premisses. To the right of each proposition will be stated the rule of derivation used to justify its appearance at that stage and (where necessary) the numbers of the premisses used. To the left of each proposition will appear the numbers of the original assumptions on which the argument at that stage depends.

Rule of Assumptions (A)

The first rule of derivation to be introduced is the *rule of assumptions*, which we call A. This rule permits us to introduce at *any* stage of an argument *any* proposition we choose as an assumption of the argument. We simply write the proposition down as a new line, write ' A ' to the right of it, and to the left of it we put its own number to show that it depends on itself as an assumption. Thus we might begin an argument

$$1 \quad (1)\ P \rightarrow Q \quad A$$

This means that our first step has been to assume the proposition $P \rightarrow Q$ by the rule of assumptions. Or after nine lines of argument we may proceed

$$10 \quad (10)\ -Q \quad A$$

This means that at the tenth line we assume the proposition $-Q$ by the rule of assumptions.

It may seem dangerously liberal that the rule of assumptions imposes no limits on the kind of assumptions we may make (in particular there is no question of ensuring that assumptions made are true). This is best understood by reminding ourselves that the logician's concern is with the soundness of the argument rather than the truth or falsity of any assumptions made; hence A allows us to make any assumptions we please—the job of the logician is to make sure that any conclusion based on them is validly based, *not* to investigate their credentials.

Modus ponendo ponens (MPP)

The second rule of derivation concerns the operator $\rightarrow$. We name it *modus ponendo ponens*, abbreviated to MPP, which was the medieval term for a closely related principle of reasoning. Given as premisses a conditional proposition and the antecedent of that conditional, MPP permits us to draw the consequent of the conditional as a conclusion. For example, given $P \rightarrow Q$ and P, we can deduce Q. Or, to take a more complicated example, given $-Q \rightarrow (-P \rightarrow Q)$

9

and $-Q$, we can deduce $-P \rightarrow Q$. Written more formally, these two arguments become:

1	1	(1) $P \rightarrow Q$	A
	2	(2) P	A
	1,2	(3) Q	1,2 MPP

2	1	(1) $-Q \rightarrow (-P \rightarrow Q)$	A
	2	(2) $-Q$	A
	1,2	(3) $-P \rightarrow Q$	1,2 MPP

On the first two lines of each of these arguments, we make the required assumptions by the rule A, numbering on the left accordingly. At line (3), we draw the appropriate conclusion by the rule MPP: the *consequent* of the conditional at line (1), given at line (2) the *antecedent* of that conditional. To the right at line (3) in both cases, we note the rule used (MPP) together with the numbers of the premises used in this application of the rule. To the left at line (3), we mark the assumptions on which the conclusion rests—in this case again (1) and (2), which here are both premises for the application of MPP and assumptions of the total argument.

Here are more complicated examples, using only the two rules A and MPP. I shall show first that, given the assumptions $P \rightarrow Q$, $Q \rightarrow R$, and P, we may validly conclude R.

3	1	(1) $P \rightarrow Q$	A
	2	(2) $Q \rightarrow R$	A
	3	(3) P	A
	1,3	(4) Q	1,3 MPP
	1,2,3	(5) R	2,4 MPP

The first three lines here merely make the necessary assumptions. At line (4), we draw by MPP the conclusion Q, given at line (1) the conditional $P \rightarrow Q$ and at line (3) its antecedent P. Hence (1) and (3) are mentioned to the right as premises for the application of the rule and to the left as the assumptions used at that stage. At line (5), we use Q, the conclusion at line (4), as a premiss for a new application of MPP, noting that Q is the antecedent of the conditional $Q \rightarrow R$ assumed at line (2). So we obtain the desired conclusion R from (2) and (4) as premises. The numbers 2 and 4 appear on the right accordingly. In deciding what assumptions to cite on the left, we note that (4) rests on (1) and (3), whilst (2) rests only on itself: we ' pool ' these assumptions to obtain (1), (2), and (3).

Secondly, I show that, given $P \rightarrow (Q \rightarrow R)$, $P \rightarrow Q$, and P, we may validly conclude R.

4	1	(1) $P \rightarrow (Q \rightarrow R)$	A
	2	(2) $P \rightarrow Q$	A
	3	(3) P	A
	1,3	(4) $Q \rightarrow R$	1,3 MPP
	2,3	(5) Q	2,3 MPP
	1,2,3	(6) R	4,5 MPP

At lines (4) and (5), the premises used for the applications of MPP are also assumptions, so that the same pair of numbers appears on the right and on the left. But at line (6), the premises are the conditional (4), $Q \rightarrow R$, and its antecedent (5), Q, neither of which are assumptions of the argument as a whole: in determining the numbers on the left, therefore, we ' pool ' the assumptions on which (4) and (5) rest—(1), (3) and (2), (3) respectively—to obtain (1), (2), and (3).

It should be obvious that MPP is a reliable principle of reasoning. It can never lead us, at least, from true premises to a false conclusion. For it is a basic feature of our use of ' if . . . then . . .' that if a conditional is true and if also its antecedent is true then its consequent must be true too, and MPP precisely allows us to affirm as a conclusion the consequent of a conditional, given as premises the conditional itself and its antecedent.

It will be a help to have an abbreviation for the cumbersome expression ' given as assumptions . . ., we may validly conclude . . .'. To this end, I introduce the symbol

$$\vdash,$$

called often but misleadingly in the literature of logic the *assertion-sign*. It may conveniently be read as ' therefore '. Before it, we list (in any order) our assumptions, and after it we write the conclusion drawn. Using this notation, we may conveniently sum up the four pieces of reasoning above (from now on to be called *proofs*) thus:

11

1 $P \rightarrow Q, P \vdash Q$;

2 $-Q \rightarrow (-P \rightarrow Q), -Q \vdash -P \rightarrow Q$;

3 $P \rightarrow Q, Q \rightarrow R, P \vdash R$;

4 $P \rightarrow (Q \rightarrow R), P \rightarrow Q, P \vdash R$.

Results obtained in this form we shall call *sequents*. Thus a sequent is an argument-frame containing a set of assumptions and a conclusion which is claimed to follow from them. Effectively, sequents which we can prove embody valid patterns of argument in the sense that, if we take the P, Q, R, . . . in a proved sequent to be actual propositions, then, reading '$\vdash$' as 'therefore', we obtain a valid argument. The propositions to the left of '$\vdash$' become assumptions of the argument, and the proposition to the right becomes a conclusion validly drawn from those assumptions. From this point of view, in constructing proofs we are demonstrating the validity of patterns of argument, which is one of the logician's chief concerns.

The sequent proved can be written down immediately from the last line of the proof.[1] In place of the numbers on the left, we write the propositions appearing on the corresponding lines; then we place the assertion sign; finally, we add as conclusion the proposition on the last line itself. To see this, the four sequents above should be compared with the last lines of the corresponding proofs.

Modus tollendo tollens (MTT)

The third rule of derivation concerns both $\rightarrow$ and $-$. Again we use a medieval term for it, *modus tollendo tollens*, abbreviated to MTT. Given as premisses a conditional proposition and *the negation* of its consequent, MTT permits us to draw *the negation* of the antecedent of the conditional as a conclusion.

Here are two simple examples of the use of MTT. I set the precedent of citing the sequent proved before the proof.

5 $P \rightarrow Q, -Q \vdash -P$

1	(1) $P \rightarrow Q$	A
2	(2) $-Q$	A
1,2	(3) $-P$	1,2 MTT

---

[1] Thus we take a proof as a proof of a *sequent*; but it is also natural to say, in a different sense, that in a proof a *conclusion* is proved *from* certain assumptions. This resultant ambiguity in the word 'prove' is fairly harmless.

6 $P \rightarrow (Q \rightarrow R)$, P, $-R \vdash -Q$

1	(1) $P \rightarrow (Q \rightarrow R)$	A
2	(2) P	A
3	(3) $-R$	A
1,2	(4) $Q \rightarrow R$	1,2 MPP
1,2,3	(5) $-Q$	3,4 MTT

For line (5), we notice that (3), $-R$, is the negation of the consequent of the conditional (4), $Q \rightarrow R$, so that by MTT we may conclude the negation $-Q$ of the antecedent of (4): to the right, we cite (3) and (4), and to the left (1) and (2)—the assumptions on which (4) rests—and (3)—the assumption, namely itself, on which (3) rests.

We may see the soundness of the rule MTT by ordinary examples. The following are evidently sound arguments:

(4) If Napoleon was Chinese, then he was Asiatic; Napoleon was not Asiatic; therefore he was not Chinese.

(5) If Napoleon was French, then he was European; Napoleon was not European; therefore he was not French.

In both cases, given a conditional and the negation of its consequent, we deduce validly the negation of its antecedent: in (4) the conclusion is true, and so are both premisses; in (5) the conclusion is false, but so is one premiss. It should be clear that this pattern of reasoning will never lead from premisses which are *all* true to a *false* conclusion.

Double negation (DN)

The fourth rule of derivation purely concerns negation. By the *double negation* of a proposition P we understand the proposition $--P$. Intuitively, to affirm that it is not the case that it is not the case that it is raining is the same as to affirm that it is raining, and this holds for any proposition whatsoever: the double negation of a proposition is identical with the proposition itself. Hence from the double negation of a proposition we can derive validly the proposition, and vice versa. This principle lies behind the *rule of double negation* (DN): given as premiss the double negation of a proposition, DN permits us to draw the proposition itself as conclusion; and given as premiss any proposition, DN permits us to draw its

13

double negation as conclusion. Unlike MPP and MTT, DN requires only one premiss for its application, not two. Its use is exemplified in the following proofs.

7 $P \rightarrow -Q$, $Q \vdash -P$

1	(1) $P \rightarrow -Q$	A
2	(2) Q	A
2	(3) $--Q$	2 DN
1,2	(4) $-P$	1,3 MTT

Note especially that, since the consequent of (1) $P \rightarrow -Q$ is $-Q$, we need to obtain the *negation* of this, i.e. $--Q$, before we can apply the rule MTT: hence we require the step of DN from (2) to (3) before the use of MTT at line (4).

8 $-P \rightarrow Q$, $-Q \vdash P$

1	(1) $-P \rightarrow Q$	A
2	(2) $-Q$	A
1,2	(3) $--P$	1,2 MTT
1,2	(4) P	3 DN

Note especially that from (1) and (2) by MTT we draw as conclusion the *negation* of the antecedent of (1), i.e. $--P$: hence we require the step of DN from (3) to (4) in order to obtain the conclusion P. Note also that the conclusion of an application of DN rests on exactly the same assumptions as its premiss.

Conditional proof (CP)

The rules MPP and MTT enable us to use a conditional *premiss*, together with either its antecedent or the negation of its consequent, in order to obtain a certain conclusion, either its consequent or the negation of its antecedent. But how may we derive a conditional *conclusion*? The most natural device is to take the antecedent of the conditional we wish to prove as an extra assumption, and aim to derive its consequent as a conclusion: if we succeed, we may take this as a proof of the original conditional from the original assumptions (if any). For example, given that all Germans are

Europeans, how might we prove that if Napoleon was German then he was European? We naturally say: suppose Napoleon was German (here we take the antecedent of the conditional to be proved as an extra assumption); now all Germans are Europeans (the given assumption); therefore Napoleon was European (here we derive the consequent as conclusion); so if Napoleon was German he was European (here we treat the previous steps of the argument as a proof of the desired conditional).

The fifth rule of derivation, the *rule of conditional proof* (CP), imitates exactly this natural procedure and is our most general device for obtaining conditional conclusions. Its working is harder to grasp than that of the earlier rules, but familiarity with it is indispensable. I first state it, then exemplify and discuss it.

Suppose some proposition (call it B) depends, as one of its assumptions, on a proposition (call it A); then CP permits us to derive the conclusion A $\rightarrow$ B on the remaining assumptions (if any). In other words, at a certain stage in a proof we have derived the conclusion B from assumption A (and perhaps other assumptions); then CP enables us to take this as a proof of A $\rightarrow$ B from the other assumptions (if any).

For example:

9 $P \rightarrow Q \vdash -Q \rightarrow -P$

1	(1) $P \rightarrow Q$	A	
2	(2) $-Q$	A	
1,2	(3) $-P$	1,2 MTT	
1	(4) $-Q \rightarrow -P$	2,3 CP	

In attempting to derive the conditional $-Q \rightarrow -P$ from $P \rightarrow Q$, we first assume its antecedent $-Q$ at line (2), and derive its consequent $-P$ at line (3); CP at line (4) enables us to treat this as a proof of $-Q \rightarrow -P$ from just assumption (1). On the right, we give first the number of the assumed antecedent and second the number of the concluded consequent. On the left, the assumption (2) at line (3) disappears into the antecedent of the new conditional, and we are left with (1) alone. Always, in an application of CP, the number of assumptions falls by one in this manner, the one omitted being called the *discharged assumption*.

10 $P \rightarrow (Q \rightarrow R) \vdash Q \rightarrow (P \rightarrow R)$

1	(1) $P \rightarrow (Q \rightarrow R)$	A
2	(2) Q	A
3	(3) P	A
1,3	(4) $Q \rightarrow R$	1,3 MPP
1,2,3	(5) R	2,4 MPP
1,2	(6) $P \rightarrow R$	3,5 CP
1	(7) $Q \rightarrow (P \rightarrow R)$	2,6 CP

A more complicated example, involving double use of CP: in attempting to derive the conditional $Q \rightarrow (P \rightarrow R)$ from $P \rightarrow (Q \rightarrow R)$, we first assume its antecedent Q at line (2), and aim to derive its consequent $P \rightarrow R$; since this consequent is also conditional, we assume *its* antecedent P at line (3), and aim to derive its consequent R. This is achieved by two steps of MPP (lines (4) and (5)); at line (6), we treat this by CP as a proof of $P \rightarrow R$ from assumptions (1) and (2), and we cite to the right line (3) (the assumption of the antecedent) and line (5) (the derivation of the consequent). In turn, we treat this at line (7) as a proof of $Q \rightarrow (P \rightarrow R)$ from assumption (1) alone, and we cite to the right line (2) (the assumption of its antecedent) and line (6) (the derivation of its consequent). As before, the assumptions on the left decrease by one at each step of CP.

11 $Q \rightarrow R \vdash (-Q \rightarrow -P) \rightarrow (P \rightarrow R)$

1	(1) $Q \rightarrow R$	A
2	(2) $-Q \rightarrow -P$	A
3	(3) P	A
3	(4) $--P$	3 DN
2,3	(5) $--Q$	2,4 MTT
2,3	(6) Q	5 DN
1,2,3	(7) R	1,6 MPP
1,2	(8) $P \rightarrow R$	3,7 CP
1	(9) $(-Q \rightarrow -P) \rightarrow (P \rightarrow R)$	2,8 CP

This proof uses all five rules of derivation introduced so far, and deserves study. Aiming to prove a complex conditional, we assume its antecedent $-Q \rightarrow -P$ at line (2), and try to prove its consequent $P \rightarrow R$. Since this is conditional, we assume its antecedent P at line (3), and after a series of steps using DN, MTT, and MPP we derive its consequent R at line (7). Two steps of CP, paralleling the last two steps in the proof of 10, complete the job by discharging in turn the assumptions (3) and (2).

Proofs 10 and 11 suggest a useful and important general method for discovering the proofs of sequents with complex conditionals as conclusion. After using the rule A for the assumptions given in the sequent, we assume also the antecedent of the desired conditional conclusion, and aim to prove its consequent; if this is also a conditional, we assume its antecedent, and aim to prove its consequent; we repeat this procedure, until our target becomes to prove a nonconditional conclusion. If we can derive this from the assumptions we now have, the right number of CP steps, applied in reverse order, will prove the original sequent.

I end this section with a remark on two common fallacies, so common that they have received names. In accordance with rule MPP, if a conditional is true and also its antecedent, then we can soundly derive its consequent. If a conditional is true and also its *consequent*, is it sound to derive its *antecedent*? The following example shows that it is not sound to do so: it is true that if Napoleon was German then he was European, since all Germans are Europeans; and it is true that Napoleon was European; but it is false, and so cannot soundly be deduced from these true premisses, that Napoleon was German. To suppose that it is sound to derive the antecedent of a conditional from the conditional and its consequent is to commit the fallacy of *affirming the consequent*. Again, in accordance with rule MTT, if a conditional is true and also the negation of its consequent, then we can soundly derive the negation of its antecedent. But it is not sound to derive the negation of a conditional's *consequent* from the conditional itself and the negation of its *antecedent*, and to suppose that it is sound is to commit the fallacy of *denying the antecedent*. The same example may be used: it is true that if Napoleon was German then he was European, and true also that he was *not* German; but it is not true that Napoleon was not European.

Put schematically: the sequents

1 $P \rightarrow Q, P \vdash Q$ and

5 $P \rightarrow Q, -Q \vdash -P$

are sound patterns of reasoning, as we have proved. But the sequents

6 $P \rightarrow Q, Q \vdash P$ and

7 $P \rightarrow Q, -P \vdash -Q$

are not sound patterns, as we have shown by finding *examples* of propositions P and Q such that the assumptions of (6) and (7) turn out true whilst their conclusions turn out false; for it is a necessary condition of a sound pattern of argument that it shall *never* lead us from assumptions that are all true to a false conclusion. (6) is in fact the pattern of the fallacy of affirming the consequent, and (7) that of the fallacy of denying the antecedent.

EXERCISES

1 Find proofs for the following sequents, using the rules of derivation introduced so far:

(a) $P \rightarrow (P \rightarrow Q), P \vdash Q$

(b) $Q \rightarrow (P \rightarrow R), -R, Q \vdash -P$

(c) $P \rightarrow --Q, P \vdash Q$

(d) $--Q \rightarrow P, -P \vdash -Q$

(e) $-P \rightarrow -Q, Q \vdash P$

(f) $P \rightarrow -Q \vdash Q \rightarrow -P$

(g) $-P \rightarrow Q \vdash -Q \rightarrow P$

(h) $-P \rightarrow -Q \vdash Q \rightarrow P$

(i) $P \rightarrow Q, Q \rightarrow R \vdash P \rightarrow R$

(j) $P \rightarrow (Q \rightarrow R) \vdash (P \rightarrow Q) \rightarrow (P \rightarrow R)$

(k) $P \rightarrow (Q \rightarrow (R \rightarrow S)) \vdash R \rightarrow (P \rightarrow (Q \rightarrow S))$

(l) $P \rightarrow Q \vdash (Q \rightarrow R) \rightarrow (P \rightarrow R)$

(m) $P \vdash (P \rightarrow Q) \rightarrow Q$

(n) $P \vdash (-(Q \rightarrow R) \rightarrow -P) \rightarrow (-R \rightarrow -Q)$

2 Show that the following sequents are unsound patterns of argument, by finding actual propositions for P and Q such that the assumption(s) are true and the conclusion false:

18

(a) $P \rightarrow -Q, -P \vdash Q$

(b) $-P \rightarrow -Q, -Q \vdash -P$

(c) $P \rightarrow Q \vdash Q \rightarrow P$

3 CONJUNCTION AND DISJUNCTION

Of the four sentence-forming operators on sentences mentioned in the last section as being of importance to the logician, only two have so far been discussed: ' if . . . then . . .' and ' not '. In the present section, we introduce rules for arguments involving '. . . and . . .' and ' either . . . or . . .'.

Let P and Q be any two propositions. Then the proposition that *both P and Q* is called the *conjunction* of P and Q and is written

$$P \& Q.$$

P and Q are called the *conjuncts* of the conjunction $P \& Q$. Similarly, the proposition that *either P or Q* is called the *disjunction* of P and Q and is written

$$P \vee Q.$$

P and Q are called the *disjuncts* of the disjunction $P \vee Q$. (The symbol ' $\vee$ ' is intended to remind classicists of the Latin ' vel ' as opposed to ' aut ': for $P \vee Q$ is understood not to exclude the possibility that both P and Q might be the case.)

There are two rules of derivation concerning &, the *rule of &-introduction* and the *rule of &-elimination*; and there are two rules concerning $\vee$, the *rule of $\vee$-introduction* and the *rule of $\vee$-elimination*. Introduction-rules serve the purpose of enabling us to derive *conclusions* containing & or $\vee$, whilst elimination-rules serve the purpose of enabling us to use *premisses* containing & or $\vee$. We discuss and exemplify these rules in turn.

&-introduction (&I)

The rule of &-introduction (&I) is exceptionally easy to master. Given any two propositions as premisses, &I permits us to derive their conjunction as a conclusion. The rule clearly corresponds to a sound principle of reasoning; for if A and B are the case separately, it is obvious that A & B must be the case. The following proofs exemplify the use of &I.

12 $P, Q \vdash P \mathbin{\&} Q$

1	(1) P	A
2	(2) Q	A
1,2	(3) $P \mathbin{\&} Q$	1.2 &I

At line (3), by &I we conclude the conjunction of the assumptions (1) and (2). To the right, we cite (1) and (2) as the premises for the application of &I; to the left we cite the pool of the assumptions on which these premises rest—in this case themselves.

13 $(P \mathbin{\&} Q) \rightarrow R \vdash P \rightarrow (Q \rightarrow R)$

1	(1) $(P \mathbin{\&} Q) \rightarrow R$	A
2	(2) P	A
3	(3) Q	A
2,3	(4) $P \mathbin{\&} Q$	2,3 &I
1,2,3	(5) R	1,4 MPP
1,2	(6) $Q \rightarrow R$	3,5 CP
1	(7) $P \rightarrow (Q \rightarrow R)$	2,6 CP

In attempting to prove the conditional $P \rightarrow (Q \rightarrow R)$, we assume first its antecedent P (line (2)) and second the antecedent of its consequent Q (line (3)). A step of &I at line (4) gives us the conjunction of these assumptions, enabling us to apply MPP at line (5) to obtain R. Two steps of CP complete the proof.

&-elimination (&E)

The rule of &-elimination (&E) is just as straightforward. Given any conjunction as premiss, &E permits us to derive either conjunct as a conclusion. Again, the rule is evidently sound; for if A & B is the case, it is obvious that A separately and B separately must be the case. Here are examples.

14 $P \mathbin{\&} Q \vdash P$

1	(1) $P \mathbin{\&} Q$	A
1	(2) P	1 &E

15 $P \& Q \vdash Q$

1	(1)	$P \& Q$	A
1	(2)	Q	1 &E

16 $P \rightarrow (Q \rightarrow R) \vdash (P \& Q) \rightarrow R$

1	(1)	$P \rightarrow (Q \rightarrow R)$	A
2	(2)	$P \& Q$	A
2	(3)	P	2 &E
2	(4)	Q	2 &E
1,2	(5)	$Q \rightarrow R$	1,3 MPP
1,2	(6)	R	4,5 MPP
1	(7)	$(P \& Q) \rightarrow R$	2,6 CP

We desire the conditional conclusion $(P \& Q) \rightarrow R$, and so we assume its antecedent at line (2) and aim for R. &E is used at lines (3) and (4) to obtain the conjuncts P and Q separately, which are required for the MPP steps at lines (5) and (6). To the right, in an application of &E, we cite the conjunction employed as a premiss, and to the left the assumptions on which that conjunction rests.

The rules &I and &E are frequently used together in the same proof. For example:

17 $P \& Q \vdash Q \& P$

1	(1)	$P \& Q$	A
1	(2)	P	1 &E
1	(3)	Q	1 &E
1	(4)	$Q \& P$	3,2 &I

18 $Q \rightarrow R \vdash (P \& Q) \rightarrow (P \& R)$

1	(1)	$Q \rightarrow R$	A
2	(2)	$P \& Q$	A
2	(3)	P	2 &E
2	(4)	Q	2 &E

1,2	(5) R	1,4 MPP
1,2	(6) P & R	3,5 &I
1	(7) $(P$ & $Q) \rightarrow (P$ & $R)$	2,6 CP

We desire the conditional conclusion $(P$ & $Q) \rightarrow (P$ & $R)$; hence we assume the antecedent P & Q and aim for P & R. This aim is translated into the aim for P and R separately, from which P & R will follow by &I. P follows from P & Q by &E, and so does Q, which can be used in conjunction with line (1) to obtain R by MPP (line (5)). When &I is used at line (6), the premisses are (3) and (5), and these rest respectively on assumption (2) and assumptions (1) and (2). Hence the pool of these—(1) and (2)—is cited to the left.

v-*introduction* (vI)

The rule of v-introduction we name vI. Given any proposition as premiss, vI permits us to derive the disjunction of that proposition and *any* proposition as a conclusion. Thus from P as premiss, we may derive P v Q as a conclusion, or Q v P as a conclusion; and here it makes no difference what proposition Q is. Clearly the conclusion will in general be much weaker than the premiss, in an application of vI. It may, that is, be the case that either P or Q even when it is not the case that P. None the less, the rule is acceptable in the sense that when P is the case it must be also the case that either P or Q. For example, it is the case that Charles I was beheaded. It follows that either he was beheaded or he was sent to the electric chair, even though of course he was not sent to the electric chair. A disjunction P v Q is true if *at least one* of its disjuncts is true, so that rule vI cannot lead from a true premiss to a false conclusion (though it may lead to a dull one).

v-*elimination* (vE)

The rule of v-elimination (vE) is rather more complex. I first state it, then explain and justify it, and finally exemplify both it and vI. Let A, B, and C be any three propositions, and suppose (*a*) that we are given that A v B, (*b*) that from A as an assumption we can derive C as conclusion, (*c*) that from B as an assumption we can derive C as conclusion ; then vE permits us to draw C as a conclusion from any assumptions on which A v B rests, together with any assumptions (apart from A itself) on which C rests in its derivation from A and any assumptions (apart from B itself) on which C rests in its derivation from B. Thus the typical situation for a step of vE is as follows: we have a disjunction A v B as a premiss, and wish to derive a certain conclusion C; we aim first to derive C from the first disjunct A, and second to derive C from the second disjunct B. When these phases of the argument are completed, we have the situation described in (*a*), (*b*), and (*c*) above, and can apply vE to obtain the conclusion C direct from A v B. On the right, we unfortunately need to cite five lines: (i) the line where the disjunction A v B appears; (ii) the line where A is assumed; (iii) the line where C is derived from A; (iv) the line where B is assumed; (v) the line where C is derived from B. And on the left, the conclusion may rest on rather a complex pool of assumptions, derived from three sources: (i) any assumptions on which A v B rests; (ii) any assumptions on which C rests in its derivation from A, though not A itself; (iii) any assumptions on which C rests in its derivation from B, though not B itself.

Though involved to state exactly, the rule vE corresponds to an entirely natural principle of reasoning. Suppose it is the case that either A or B, i.e. that *one* of A or B is true; and suppose that on the assumption A, we can show C to be the case, i.e. that if A holds C holds; suppose also that on the assumption B, we can still show that C holds, i.e. that if B holds C also holds; then C holds either way. For example: you agree that either it is raining or it is fine (A v B); given that it is raining, then it is not fit to go for a walk (from A we derive C); given that it is fine, then it must be very hot, so that again it is not fit to go for a walk (from B we derive C). Hence either way it is not fit to go for a walk (we conclude C).

19 *P* v *Q* ⊢ *Q* v *P*

1	(1) *P* v *Q*	A
2	(2) *P*	A
2	(3) *Q* v *P*	2 vI
4	(4) *Q*	A
4	(5) *Q* v *P*	4 vI
1	(6) *Q* v *P*	1,2,3,4,5 vE

On line (1), we assume $P \vee Q$; since this is a disjunction, we aim to derive the conclusion $Q \vee P$ from the first disjunct P, assumed at line (2), and also from the second disjunct Q, assumed at line (4). This is achieved on lines (3) and (5) by steps of vI which should be obvious. At line (6), we conclude $Q \vee P$ from assumption (1) directly, since it follows from each disjunct separately. On the right, we cite line (1) (the disjunction), line (2) (assumption of first disjunct), line (3) (derivation of conclusion from that disjunct), line (4) (assumption of second disjunct), and line (5) (derivation of conclusion from that disjunct). To the left, we cite any assumptions on which the disjunction rests (here (1) rests on itself, which is therefore cited), together with any assumptions used to derive the conclusion from the disjuncts apart from the disjuncts themselves (inspection of the citations to the left of line (3) and (5) shows that there are none such). This proof should reveal the importance of keeping accurate assumption-records on the left of proofs: lines (3) and (5) here give indeed the right conclusion $Q \vee P$, but not from the right assumption, which is (1); this is achieved only at line (6), which differs from lines (3) and (5) in the annotation on the left.

20 $Q \rightarrow R \vdash (P \vee Q) \rightarrow (P \vee R)$

1	(1) $Q \rightarrow R$	A
2	(2) $P \vee Q$	A
3	(3) P	A
3	(4) $P \vee R$	3 vI
5	(5) Q	A
1,5	(6) R	1,5 MPP
1,5	(7) $P \vee R$	6 vI
1,2	(8) $P \vee R$	2,3,4,5,7 vE
1	(9) $(P \vee Q) \rightarrow (P \vee R)$	2,8 CP

The desired conclusion here is conditional; so we assume its antecedent $P \vee Q$ (line (2)), and aim to derive $P \vee R$; this assumption is a disjunction, so we assume each disjunct in turn (lines (3) and (5)) and derive the conclusion $P \vee R$ from each (lines (4) and (7)). Hence the citation on the right at line (8) is 2,3,4,5,7. The assumptions at line (8) are those on which the disjunction $P \vee Q$ rests (itself, (2)), together with any used to obtain $P \vee R$ from (3) apart from (3) itself (none, as line (4) reveals) and any used to obtain $P \vee R$ from (5) apart from (5) itself (namely (1), as line (7) reveals). A step of CP completes the proof from (1) of the desired conditional.

21 $P \vee (Q \vee R) \vdash Q \vee (P \vee R)$

1	(1) $P \vee (Q \vee R)$	A	
2	(2) P	A	
2	(3) $P \vee R$	2 vI	
2	(4) $Q \vee (P \vee R)$	3 vI	
5	(5) $Q \vee R$	A	
6	(6) Q	A	
6	(7) $Q \vee (P \vee R)$	6 vI	
8	(8) R	A	
8	(9) $P \vee R$	8 vI	
8	(10) $Q \vee (P \vee R)$	9 vI	
5	(11) $Q \vee (P \vee R)$	5,6,7,8,10 vE	
1	(12) $Q \vee (P \vee R)$	1,2,4,5,11 vE	

This proof deserves detailed study, in the use both of vI and of vE. Careful attention to bracketing is required. The assumption is a disjunction, the second of whose disjuncts is a disjunction itself. The proof falls into two distinct parts, lines (2)–(4) and lines (5)–(11): the first part establishes the desired conclusion from the first disjunct of the original disjunction (line (4)), and the second part establishes the same conclusion from the second disjunct (line (11)). This should explain the final step of vE at line (12). The second part (lines (5)–(11)), which begins with a disjunctive assumption, also falls into two sub-parts and involves a subsidiary step of vE at line (11). Lines (6)–(7) obtain the conclusion from the first disjunct Q of (5), and lines (8)–(10) obtain the conclusion from its second disjunct R. Hence the final conclusion is obtained no less than five times in the proof, from different assumptions each time.

Reductio ad absurdum (RAA)

The last rule to be introduced at this stage is in many ways the most powerful and the most useful; it is easy to understand, though a little difficult to state precisely. We shall call it the *rule of reductio ad absurdum* (RAA). First, we define a *contradiction*. A *contradiction* is a conjunction the second conjunct of which is the negation of the first conjunct: thus $P \& -P$, $R \& -R$, $(P \rightarrow Q) \& -(P \rightarrow Q)$ are all contradictions. Now suppose that from an assumption A, together perhaps with other assumptions, we can derive a contradiction as a conclusion; then RAA permits us to derive $-A$ as a conclusion from those other assumptions (if any). This rule rests on the natural principle that, if a contradiction can be deduced from a proposition A, A cannot be true, so that we are entitled to affirm its negation $-A$.

Here are examples.

22 $P \rightarrow Q, P \rightarrow -Q \vdash -P$

1	(1) $P \rightarrow Q$	A
2	(2) $P \rightarrow -Q$	A
3	(3) P	A
1,3	(4) Q	1,3 MPP
2,3	(5) $-Q$	2,3 MPP
1,2,3	(6) $Q \& -Q$	4,5 &I
1,2	(7) $-P$	3,6 RAA

This is a typical example of the use of RAA. Aiming at the conclusion $-P$, we assume (line (3)) P and hope to derive from it a contradiction; for, if P leads to a contradiction, we can conclude $-P$ by RAA. We obtain the contradiction $Q \& -Q$ at line (6), and so conclude $-P$ at line (7). On the right, we cite the assumption which we are blaming for the contradiction—the one whose negation we conclude in the RAA step, here (3)—and the contradiction itself, here (6). On the left, as in a CP step, the number of assumptions naturally falls by one, there being omitted the one which we blame for the contradiction.

23 $P \rightarrow -P \vdash -P$

1	(1)	$P \rightarrow -P$	A
2	(2)	P	A
1,2	(3)	$-P$	1,2 MPP
1,2	(4)	$P \& -P$	2,3 &I
1	(5)	$-P$	2,4 RAA

Again desiring $-P$, we assume P (line (2)) and obtain a contradiction (line (4)). Given (1), therefore, we conclude $-P$ by RAA. The sequent proved is striking, and perhaps unexpected—given that if a proposition is the case then so is its negation, we can conclude that its negation is true. This is the first surprising result to be established by our rules, but there will be more.

The rule RAA is particularly useful when we wish to derive *negative* conclusions. It suggests that, instead of attempting a direct proof, we should assume the corresponding *affirmative* proposition and aim to derive a contradiction, thus indirectly establishing the negative. It can also be used, however, to establish affirmatives themselves, *via* DN. If we want to derive A, we may assume $-A$ and obtain a contradiction. Hence by RAA we can conclude $--A$ (the negation of what we assumed) and so by DN we obtain A. It is a good general tip for proof-discovery that, when direct attempts fail, often an RAA proof will succeed.

So far, ten rules of derivation have been introduced: we shall need no new ones until Chapter 3.

EXERCISES

1   Find proofs for the following sequents:

(a) $P \vdash Q \rightarrow (P \& Q)$

(b) $P \& (Q \& R) \vdash Q \& (P \& R)$

(c) $(P \rightarrow Q) \& (P \rightarrow R) \vdash P \rightarrow (Q \& R)$

(d) $Q \vdash P \vee Q$

(e) $P \& Q \vdash P \vee Q$

(f) $(P \rightarrow R) \& (Q \rightarrow R) \vdash (P \vee Q) \rightarrow R$

(g) $P \rightarrow Q, R \rightarrow S \vdash (P \& R) \rightarrow (Q \& S)$

(h) $P \rightarrow Q, R \rightarrow S \vdash (P \vee R) \rightarrow (Q \vee S)$

(i) $P \rightarrow (Q \& R) \vdash (P \rightarrow Q) \& (P \rightarrow R)$

(j) $-P \rightarrow P \vdash P$

2 Show that the following sequents are unsound, by finding actual propositions for P and Q such that the assumption is true and the conclusion false:

(a) $P \vdash P \& Q$

(b) $P \lor Q \vdash P$

(c) $P \lor Q \vdash P \& Q$

(d) $P \rightarrow Q \vdash P \& Q$

4 THE BICONDITIONAL

There is a sentence-forming operator on sentences, of considerable importance to the logician though of rare occurrence in ordinary speech, which we have not so far introduced. This is '. . . if and only if . . .'. We study it in the present section.

To begin with, let us consider the differences between ' if . . . then . . .' and ' only if . . . then . . .'. Compare the following two propositions:

(1) if it snows it turns colder;
(2) only if it snows it turns colder.

(1) affirms that its snowing is *sufficient* for it to turn colder, whilst (2) affirms that its snowing is *necessary* for it to turn colder, that if it is to turn colder it *must* snow. Hence we shall say that, whenever it is the case that if P then Q, P is a *sufficient condition* for Q, and, whenever it is the case that only if P then Q, P is a *necessary condition* for Q. To make this fundamental distinction clearer, let us compare

(3) if you hit the glass with a hammer, you will break it;
(4) only if you hit the glass with a hammer will you break it.

(3) is very likely true; (4) is very likely false, since there are other ways of breaking the glass than by wielding a hammer. On the other hand, of the two propositions

(5) if you use a screwdriver, you will unscrew that very tight screw;
(6) only if you use a screwdriver will you unscrew that very tight screw,

(5) may very well be false (you may use the screwdriver and still not unscrew the screw), and (6) true, since there may be no other way of turning the screw than by wielding a screwdriver. Hitting with a hammer is (probably) a sufficient but not a necessary condition for breaking a glass; using a screwdriver is (possibly) a necessary but not a sufficient condition for turning a tight screw.

In scientific and in mathematical reasoning, and consequently in logic, we are often interested in a condition being *both sufficient and necessary*. *P* will be a sufficient and necessary condition for *Q* in just the case that *Q* holds *if and only if P* holds. Hence our interest in '. . . if and only if . . .'. It may seem, therefore, that we require a special symbol for ' only if . . . then . . .'; but that this is not so may be seen as follows.

Suppose that only if *P* then *Q*; then *P* is a necessary condition for *Q*, that is, for *Q* to be the case *P* must be the case; hence if *Q* is the case, so is *P*. For example, suppose, as before, that using a screwdriver is a necessary condition for turning the screw; then if the screw is turned, a screwdriver has been used. In short, given that only if *P* then *Q*, we can infer that if *Q* then *P*. Conversely, suppose that if *Q* then *P*; then for *Q* to be the case *P* must be the case, for if *Q* is the case and *P* not the case it cannot hold that if *Q* then *P*; hence *P* is a necessary condition for *Q*, that is, only if *P* then *Q*. These two arguments suggest that to affirm only if *P* then *Q* is to affirm if *Q* then *P*. Hence to express symbolically ' only if *P* then *Q* ' we may use ' $\rightarrow$ ' and simply write

$$Q \rightarrow P.$$

To affirm, therefore, that *Q if and only if P* is to affirm that if *P* then *Q* and only if *P* then *Q*, which is to affirm that if *P* then *Q* and if *Q* then *P*; or, in symbols,

$$(P \rightarrow Q) \& (Q \rightarrow P).$$

But, rather than use this complex expression, we may conveniently adopt a double arrow and write as an abbreviation

$$P \leftrightarrow Q.$$

(This symbol helps to emphasize the mutuality of the relationship between *P* and *Q*.) We call the proposition *P* $\leftrightarrow$ *Q* the *biconditional* of *P* and *Q*.

What are the properties of the biconditional in argument? We could lay down rules of derivation for this operator, as we have done for the four operators of the previous two sections. But in fact the properties of ' $\leftrightarrow$ ' follow readily from those of ' & ' and ' $\rightarrow$ ', in terms of which we have just defined the biconditional. For example:

24 $P \leftrightarrow Q \vdash Q \leftrightarrow P$

1	(1) $P \leftrightarrow Q$	A
1	(2) $(P \rightarrow Q) \& (Q \rightarrow P)$	1
1	(3) $P \rightarrow Q$	2 &E
1	(4) $Q \rightarrow P$	2 &E
1	(5) $(Q \rightarrow P) \& (P \rightarrow Q)$	4,3 &I
1	(6) $Q \leftrightarrow P$	5

Here the step from (1) to (2) is justified by our taking ' $P \leftrightarrow Q$ ' as an abbreviation for ' $(P \rightarrow Q) \& (Q \rightarrow P)$ ': at (2) we merely expand what we have assumed at (1). Similarly, but in reverse, the step from (5) to (6) is justified: for (6) is merely an abbreviation for the conclusion (5). However, we need to ratify such steps more precisely, and to this end we introduce the following formal definition of the biconditional:

$$Df. \leftrightarrow : \quad A \leftrightarrow B = (A \rightarrow B) \& (B \rightarrow A).$$

This definition is to be understood as a very condensed way of saying: given any two sentences A and B, we may replace in a proof the sentence A $\leftrightarrow$ B by the sentence (A $\rightarrow$ B) & (B $\rightarrow$ A), and vice versa. When this definition is applied, we shall cite ' $Df. \leftrightarrow$ ' on the right. Lines (2) and (6) of the last proof should in fact be so marked.

The next few proofs exemplify the use of this definition.

25 $P, P \leftrightarrow Q \vdash Q$

1	(1) P	A
2	(2) $P \leftrightarrow Q$	A
2	(3) $(P \rightarrow Q) \& (Q \rightarrow P)$	2 $Df. \leftrightarrow$
2	(4) $P \rightarrow Q$	3 &E
1,2	(5) Q	1,4 MPP

26 $P \leftrightarrow Q, Q \leftrightarrow R \vdash P \leftrightarrow R$

1	(1) $P \leftrightarrow Q$	A
2	(2) $Q \leftrightarrow R$	A
1	(3) $(P \to Q) \& (Q \to P)$	1 *Df.* $\leftrightarrow$
1	(4) $P \to Q$	3 &E
1	(5) $Q \to P$	3 &E
2	(6) $(Q \to R) \& (R \to Q)$	2 *Df.* $\leftrightarrow$
2	(7) $Q \to R$	6 &E
2	(8) $R \to Q$	6 &E
9	(9) P	A
1,9	(10) Q	4,9 MPP
1,2,9	(11) R	7,10 MPP
1,2	(12) $P \to R$	9,11 CP
13	(13) R	A
2,13	(14) Q	8,13 MPP
1,2,13	(15) P	5,14 MPP
1,2	(16) $R \to P$	13,15 CP
1,2	(17) $(P \to R) \& (R \to P)$	12,16 &I
1,2	(18) $P \leftrightarrow R$	17 *Df.* $\leftrightarrow$

To derive $P \leftrightarrow R$ by *Df.* $\leftrightarrow$ we need to derive $(P \to R)$ & $(R \to P)$, and we aim at each conjunct separately. The first eight lines of the proof merely itemize the information in the assumptions, by applying *Df.* $\leftrightarrow$ and &E. This information (lines (4), (5), (7), (8)) is then used in a straightforward manner to derive the two required conjuncts (lines (9)–(12) and (13)–(16)).

27 $(P \& Q) \leftrightarrow P \vdash P \to Q$

1	(1) $(P \& Q) \leftrightarrow P$	A
1	(2) $((P \& Q) \to P) \& (P \to (P \& Q))$	1 *Df.* $\leftrightarrow$
1	(3) $P \to (P \& Q)$	2 &E
4	(4) P	A
1,4	(5) $P \& Q$	3,4 MPP

1,4	(6) Q	5 &E
1	(7) $P \rightarrow Q$	4,6 CP

By $Df \leftrightarrow$, (1) is an abbreviation for (2), if we take A to be the sentence '$P \& Q$' and B to be the sentence 'P'.

28 $P \& (P \leftrightarrow Q) \vdash P \& Q$

1	(1) $P \& (P \leftrightarrow Q)$	A
1	(2) $P \& ((P \rightarrow Q) \& (Q \rightarrow P))$	1 $Df. \leftrightarrow$
1	(3) P	2 &E
1	(4) $(P \rightarrow Q) \& (Q \rightarrow P)$	2 &E
1	(5) $P \rightarrow Q$	4 &E
1	(6) Q	3,5 MPP
1	(7) $P \& Q$	3,6 &I

In the proofs preceding this, when $Df. \leftrightarrow$ was applied, it was applied to a sentence as a whole, i.e. the sentence to which it was applied was of the form $A \leftrightarrow B$; but this is not essential: here in fact at line (2) it is applied to the second conjunct of the proposition at line (1).

Although $Df. \leftrightarrow$ is like the ten rules of derivation introduced so far, in that it justifies transitions in a proof, it should not be thought of as another rule on a par with the rest. Its role in proofs is to enable us to take advantage of a piece of symbolic shorthand, rather than to enable us genuinely to derive conclusions from premisses. It happens that, for certain ends, we are interested in complex propositions such as $(P \rightarrow Q) \& (Q \rightarrow P)$, and to facilitate our study of them we agree to abbreviate our expressions for them to sentences such as '$P \leftrightarrow Q$'. This is a guide to the eye, a sop thrown to human weakness: were we brave enough, in place of 27 above, for example, we might merely prove

(7) $((P \& Q) \rightarrow P) \& (P \rightarrow (P \& Q)) \vdash P \rightarrow Q$;

but the expression we have used discloses a pattern which we might miss in the expression of (7). Given, therefore, that we wish to take advantage of this abbreviation in proofs, we need a device for transforming sentences containing '$\leftrightarrow$' into sentences lacking it, and a reverse device for transforming sentences of the right form lacking ' $\longleftrightarrow$ ' into sentences containing it: that is exactly what *Df.* $\longleftrightarrow$ provides. To put the point in a slightly different way, any logical properties which ' $\longleftrightarrow$ ' may seem to have are merely properties of ' & ' and ' $\rightarrow$ ' in symbolic disguise.

A definition such as *Df.* $\longleftrightarrow$ may be called a *stipulative* definition, in that it stipulates or lays down the meaning of the symbol ' $\longleftrightarrow$ ' in terms of symbols ' $\rightarrow$ ' and ' & ' whose meaning is known from the rules governing their deployment in proofs. To say that a definition is stipulative is not to say that it is *arbitrary* (though the actual symbol ' $\longleftrightarrow$ ' chosen is in a sense arbitrarily chosen). Indeed, I carefully prepared the ground for the definition by arguing that what we in fact understood by the proposition that *Q* if and only if *P* was that if *P* then *Q* and if *Q* then *P*. But formally the definition is stipulative in that it announces that a sign *is to be taken* in a certain way.

EXERCISES

1 Using *Df.* $\longleftrightarrow$ in conjunction with the rules of derivation of sections 2 and 3, find proofs for the following sequents:

(a) $Q, P \longleftrightarrow Q \vdash P$

(b) $P \rightarrow Q, Q \rightarrow P \vdash P \longleftrightarrow Q$

(c) $P \longleftrightarrow Q \vdash -P \longleftrightarrow -Q$

(d) $-P \longleftrightarrow -Q \vdash P \longleftrightarrow Q$

(e) $(P \vee Q) \longleftrightarrow P \vdash Q \rightarrow P$

(f) $P \longleftrightarrow -Q, Q \longleftrightarrow -R \vdash P \longleftrightarrow R$

2 Just as '. . . if and only if . . .' can be defined in terms of ' if . . . then . . .' and '. . . and . . .', so ' unless . . ., then . . .' can be defined in terms of ' if . . . then . . .' and ' not '. For to affirm that unless *P* then *Q* is to affirm that if not *P* then *Q* (justify this by taking cases). Let us, therefore, stipulate

$$Df. * : A * B = -A \rightarrow B .$$

Using *Df. ** in a way parallel to *Df.* $\longleftrightarrow$, find proofs for the following sequents:

(a) $P * Q \vdash Q * P$

(b) $P * Q, P * R \vdash P * (Q \& R)$

(c) $P * Q, R * -Q \vdash P * R$

(d) $P * P \vdash P$

(e) $-P * R, -Q * R, P \vee Q \vdash R$

5 FURTHER PROOFS: RÉSUMÉ OF RULES

We now exemplify the use of our rules of derivation by some more advanced proofs. The sequents proved are themselves worth studying, as exhibiting some of the more basic formal properties of the operators concerned; the formal work deserves attention too, since it frequently illustrates a technique which the student should master as an aid to his own discovery of proofs. After many proofs I add notes which pick out interesting features and try to indicate how the proofs are discovered. There are not, it should be remembered, precise rules for proof-discovery; hints can be given, but actual practice is all-important. (To this end, the student might try to rediscover proofs of sequents in preceding sections.) At the end of the section, I add for reference purposes a statement of the rules introduced so far.

29 $P \vdash P$

$$1 \quad (1)\, P \quad A$$

No shorter sequent than this can be proved, and its proof is the shortest possible proof: yet it is worth close attention. Line (1) affirms that, given (1), P follows; what is (1)?—the proposition P itself. That is, given P, we may conclude that P, which is the sequent to be proved. Is this really sound? It is often thought that to infer P from P is unsound, on the grounds that the argument is circular, but this is a misunderstanding; certainly the argument is circular (in the popular sense), but a circular argument is entirely sound (though extremely dull). Given that it is raining, the *safest possible* conclusion is that it is raining. If I infer a proposition from itself, I do not err in reasoning, though I do not advance in information either. From this standpoint, the rule of assumptions is precisely based on the principle of the soundness of a circular argument; for the rule of assumptions affirms that, given a certain proposition, we can at least infer that proposition.

Let A and B be two propositions such that we can prove both the sequent $A \vdash B$ and the sequent $B \vdash A$; then we say that A and B are *interderivable*, and we write the fact thus:

$$A \dashv\vdash B,$$

using a suggestive symbol. For example, a comparison of sequents 13 and 16 (Section 3) reveals that $(P \,\&\, Q) \twoheadrightarrow R$ and $P \twoheadrightarrow (Q \twoheadrightarrow R)$ are interderivable, so that we may write in summary:

30 $(P \,\&\, Q) \twoheadrightarrow R \dashv\vdash P \twoheadrightarrow (Q \twoheadrightarrow R)$

In establishing an interderivability result, the work naturally falls into two halves. Thus:

31 $P \,\&\, (P \vee Q) \dashv\vdash P$

    (a) $P \,\&\, (P \vee Q) \vdash P$

1	(1)	$P \,\&\, (P \vee Q)$	A
1	(2)	P	1 &E

    (b) $P \vdash P \,\&\, (P \vee Q)$

1	(1)	P	A
1	(2)	$P \vee Q$	1 vI
1	(3)	$P \,\&\, (P \vee Q)$	1,2 &I

In proving (31(b)) that $P \,\&\, (P \vee Q)$ follows from P, we prove that each conjunct follows separately: that P follows from P is in fact given at line (1) (compare 29 above and the note following).

32 $P \vee (P \,\&\, Q) \dashv\vdash P$

    (a) $P \vee (P \,\&\, Q) \vdash P$

1	(1)	$P \vee (P \,\&\, Q)$	A
2	(2)	P	A
3	(3)	$P \,\&\, Q$	A
3	(4)	P	3 &E
1	(5)	P	1,2,2,3,4 vE

    (b) $P \vdash P \vee (P \,\&\, Q)$

1	(1)	P	A
1	(2)	$P \vee (P \,\&\, Q)$	1 vI

In 32(a), to show that P follows from the disjunction $P \vee (P \,\&\, Q)$, we need to show that it follows from each disjunct in turn in order to apply vE. That P follows from P is given by line (2) (compare again 29), and that it follows from P & Q is proved by &E at line (4). This should explain the double citation of '2' at line (5) on the right: the first '2' signifies the assumption of the first disjunct P at that line, and the second '2' signifies that the conclusion P is derived from that assumption at the same line.

33 $P \vee P \dashv\vdash P$

    (*a*) $P \vee P \vdash P$

      1   (1) $P \vee P$ A

      2   (2) P A

      1   (3) P 1,2,2,2,2 vE

    (*b*) $P \vdash P \vee P$

      1   (1) P A

      1   (2) $P \vee P$ 1 vI

33(*a*) line (3) reveals a limiting case of the use of vE. To derive P from $P \vee P$, by vE we need to show that P follows from each disjunct in turn; but the disjuncts are the same, P itself, so that the whole work is done by line (2): hence the four citations of '2' to the right at line (3).

34 $P, -(P \,\&\, Q) \vdash -Q$

1	(1) P	A
2	(2) $-(P \,\&\, Q)$	A
3	(3) Q	A
1,3	(4) $P \,\&\, Q$	1,3 &I
1,2,3	(5) $(P \,\&\, Q) \,\&\, -(P \,\&\, Q)$	2,4 &I
1,2	(6) $-Q$	3,5 RAA

To derive $-Q$, we proceed indirectly and assume Q, hoping to obtain a contradiction; this is achieved at line (5), whence RAA yields the desired sequent. The principle of reasoning associated with 34 has the medieval name *modus ponendo tollens*: if P is the case, and it is not the case that both P and Q, then it is not the case that Q.

35 $P \rightarrow Q \dashv\vdash -(P \& -Q)$

(a) $P \rightarrow Q \vdash -(P \& -Q)$

1	(1) $P \rightarrow Q$	A
2	(2) $P \& -Q$	A
2	(3) P	2 &E
2	(4) $-Q$	2 &E
1,2	(5) Q	1,3 MPP
1,2	(6) $Q \& -Q$	4,5 &I
1	(7) $-(P \& -Q)$	2,6 RAA

(b) $-(P \& -Q) \vdash P \rightarrow Q$

1	(1) $-(P \& -Q)$	A
2	(2) P	A
3	(3) $-Q$	A
2,3	(4) $P \& -Q$	2,3 &I
1,2,3	(5) $(P \& -Q) \& -(P \& -Q)$	1,4 &I
1,2	(6) $--Q$	3,5 RAA
1,2	(7) Q	6 DN
1	(8) $P \rightarrow Q$	2,7 CP

35(a): another indirect proof—we assume (line (2)) $P \& -Q$ and aim for a contradiction. Lines (3) and (4) unpack by &E the information of line (2), and the desired contradiction is almost immediate (line (6)). 36(b) is a little more complex. Aiming to prove $P \rightarrow Q$, we assume P (line (2)) and take Q as a subsidiary target, relying on CP to redress the balance at the last step. There seems to be no direct way of deriving Q from (1) and (2), so we assume $-Q$ (line (3)) and aim for a contradiction. By &I, assumptions (2) and (3) contradict assumption (1), as we establish at line (5). By RAA, this yields $--Q$ from (1) and (2), and hence Q (line (8)) by DN.

36 $P \lor Q \dashv\vdash -(-P \& -Q)$

(a) $P \lor Q \vdash --(-P \& -Q)$

1	(1) $P \lor Q$	A

37

2	(2) $-P \& -Q$	A
3	(3) P	A
2	(4) $-P$	2 &E
2,3	(5) $P \& -P$	3,4 &I
3	(6) $-(-P \& -Q)$	2,5 RAA
7	(7) Q	A
2	(8) $-Q$	2 &E
2,7	(9) $Q \& -Q$	7,8 &I
7	(10) $-(-P \& -Q)$	2,9 RAA
1	(11) $-(-P \& -Q)$	1,3,6,7,10 vE

(*b*) $-(-P \& -Q) \vdash P \vee Q$

1	(1) $-(-P \& -Q)$	A
2	(2) $-(P \vee Q)$	A
3	(3) P	A
3	(4) $P \vee Q$	3 vI
2,3	(5) $(P \vee Q) \& -(P \vee Q)$	2,4 &I
2	(6) $-P$	3,5 RAA
7	(7) Q	A
7	(8) $P \vee Q$	7 vI
2,7	(9) $(P \vee Q) \& -(P \vee Q)$	2,8 &I
2	(10) $-Q$	7,9 RAA
2	(11) $-P \& -Q$	6,10 &I
1,2	(12) $(-P \& -Q) \& -(-P \& -Q)$	1,11 &I
1	(13) $--(P \vee Q)$	2,12 RAA
1	(14) $P \vee Q$	13 DN

Both 36(*a*) and 36(*b*) are instructive proofs, and merit close scrutiny. The basic idea of 36(*a*) is proof by vE. Given a disjunctive assumption, we assume (line (3)) the first disjunct and aim for the conclusion, and assume (line (7)) the second disjunct and aim for the same conclusion. In each case, the conclusion is obtained by RAA, so that we assume once and for all (line (2)) $-P \& -Q$ whose negation we wish to derive. Lines (3)–(6) achieve the first objective, lines

(7)–(10) the second. The basic idea of 36(*b*) is proof by RAA. We assume (line (2)) the negation of the desired conclusion, and aim for a contradiction. Clearly what contradicts assumption (1) is $-P$ & $-Q$, so that the objective becomes to derive $-P$ and $-Q$ separately from (2). To derive $-P$, we assume P (line (3)), and obtain a contradiction (line (5)); hence $-P$ follows from (2) (line (6)). In a parallel way, $-Q$ also follows from (2) (line (10)). We thus achieve the desired contradiction at line (12). It is worth noting that at line (11) of this proof we have actually proved the sequent $-(P \vee Q) \vdash -P$ & $-Q$ (compare Exercise 1 (*f*) at the end of the section).

The ten rules we have used hitherto enable us to prove interesting, and in certain cases unobvious, results concerning the interrelations of our sentence-forming operators on sentences. Yet they are all rules which after reflection we are inclined to accept as corresponding to sound and obvious principles of reasoning: at least, from true premisses we shall not be led by them to false conclusions. It should be clear by now that any insights we have so far obtained into the proper codification of arguments are mainly due to the adoption of a special logical notation and of rules the application of which can be *mechanically checked*. Indeed, if someone queries our conclusions, we can present him with the proofs and ask him to state exactly which step he regards as invalid and why. In this respect, the situation is like that in arithmetic: it is idle merely to disagree with a certain calculation; you should say *where* the mistake has been made, and why you consider it to be such. There is a difference, however: calculations can be *performed*, as well as checked, mechanically, whilst we so far know of no mechanical way of generating proofs—though, once discovered, a machine could certify them as valid.

SUMMARY OF RULES OF DERIVATION

1 *Rule of Assumptions* (A)

Any proposition may be introduced at any stage of a proof. We write to the left the number of the line itself.

2 *Modus Ponendo Ponens* (MPP)

Given A and A → B, we may derive B as conclusion. B depends on any assumptions on which either A or A → B depends.

39

3 *Modus Tollendo Tollens* (MTT)

Given $-B$ and $A \rightarrow B$, we may derive $-A$ as conclusion. $-A$ depends on any assumptions on which either $-B$ or $A \rightarrow B$ depends.

4 *Double Negation* (DN)

Given A, we may derive $--A$ as conclusion, and vice versa. In either case, the conclusion depends on the same assumptions as the premiss.

5 *Conditional Proof* (CP)

Given a proof of B from A as assumption, we may derive $A \rightarrow B$ as conclusion on the remaining assumptions (if any).

6 *&-Introduction* (&I)

Given A and B, we may derive A & B as conclusion. A & B depends on any assumptions on which either A or B depends.

7 *&-Elimination* (&E)

Given A & B, we may derive either A or B separately. In either case, the conclusion depends on the same assumptions as the premiss.

8 v-*Introduction* (vI)

Given either A or B separately, we may derive A v B as conclusion. In either case, the conclusion depends on the same assumptions as the premiss.

9 v-*Elimination* (vE)

Given A v B, together with a proof of C from A as assumption and a proof of C from B as assumption, we may derive C as conclusion. C depends on any assumptions on which A v B depends or on which C depends in its derivation from A (apart from A) or on which C depends in its derivation from B (apart from B).

10 *Reductio ad Absurdum* (RAA)

Given a proof of B & $-B$ from A as assumption, we may derive $-A$ as conclusion on the remaining assumptions (if any).

Note: The biconditional-sign ' $\leftrightarrow$ ' is introduced by the following definition:

$$Df. \leftrightarrow : \quad A \leftrightarrow B = (A \rightarrow B) \ \& \ (B \rightarrow A)$$

This definition permits the replacement of $A \leftrightarrow B$ appearing in a conclusion by $(A \rightarrow B) \ \& \ (B \rightarrow A)$, and vice versa.

EXERCISE

1   Find proofs for the following sequents:

(a) $P \vee Q \quad P \vee Q$

(b) $P \mathbin{\&} P \dashv\vdash P$

(c) $P \mathbin{\&} (Q \vee R) \dashv\vdash (P \mathbin{\&} Q) \vee (P \mathbin{\&} R)$

(d) $P \vee (Q \mathbin{\&} R) \dashv\vdash (P \vee Q) \mathbin{\&} (P \vee R)$

(e) $P \mathbin{\&} Q \dashv\vdash -(P \to -Q)$

(f) $-(P \vee Q) \dashv\vdash -P \mathbin{\&} -Q$

(g) $-(P \mathbin{\&} Q) \dashv\vdash -P \vee -Q$

(h) $P \mathbin{\&} Q \dashv\vdash -(-P \vee -Q)$

(i) $P \to Q \vdash -P \vee Q$

(j) $-P \to Q \vdash P \vee Q$

CHAPTER 2

The Propositional Calculus 2

INTRODUCTION

In the previous chapter we gradually learned what may be described as a *formal language*, a language designed for the study of certain patterns of argument in something of the way in which the language of elementary mathematics is designed for the study of certain numerical operations (addition, subtraction, etc.). This language is often called, for reasons which should be obvious, the *propositional calculus* (also sometimes the *sentential* calculus). In the present chapter, we study it at a more theoretical level, in order to gain a clearer insight into its properties and its power. Among the questions we shall raise are the following three. (i) It commonly happens in mathematics that a result, once proved, can be utilized without re-proof in obtaining new results—mathematics is *progressive* in just this sense, as any student of Euclidean geometry knows. Are there any analogous devices whereby we can use a sequent already proved to facilitate the discovery of proofs for other sequents? An affirmative answer is given in Section 2. (ii) However confident on intuitive grounds we may be that our rules of derivation are safe, is there nevertheless any way of *showing* that they are safe, showing that they will not yield sequents which are in fact invalid? A way is found in Sections 3 and 4. (iii) We have so far introduced ten rules of derivation for operating the symbols of the language: are these enough, or do we require more? Section 5 shows that our rules form in a certain sense a *complete* set, and that no more are needed. The answers to these and related questions afford a deepened understanding of the nature of the propositional calculus.

1 FORMATION RULES

The propositional calculus is, I have said, a kind of language, and as such it has a grammar or, more particularly, a *syntax*. We have taken this syntax for granted in our fairly easy-going approach so far;

but we cannot go much further without a more scrupulous account of the structure of the language itself. In particular, we have taken for granted what was understood by a *sentence in the symbolism*; it is part of our task as logicians to make this notion precise, and we devote this section to the job by introducing a rather long series of definitions.

First, I define a *bracket*. A bracket is one of the marks:

$$\text{‘ (’, ‘) ’,}$$

and I call the first kind of mark a *left-hand* bracket and the second a *right-hand* bracket. This definition, which should be readily understood, is an *ostensive* definition, so-called because I *show* or *exhibit* what a bracket is rather than use other *words* to define one. (We could avoid ostensive definition: I might say that a bracket is an arc of a circle, with one end point placed vertically above the other end point.)

Second, I define a *logical connective*, often just called a *connective*. A logical connective is one of the marks:

$$\text{‘ → ’, ‘ − ’, ‘ \& ’, ‘ v ’, ‘ ↔ ’.}$$

This is also an ostensive definition, which formally introduces the symbols employed in the last chapter for sentence-forming operators on sentences.

Third, I define a *(propositional) variable*. A propositional variable is one of the marks:

$$\text{‘ }P\text{ ’, ‘ }Q\text{ ’, ‘ }R\text{ ’, ...}$$

This is again an ostensive definition, but importantly different from the earlier two. There are just two kinds of mark which are called brackets, and just five which are called connectives; but the ‘. . .’ in the definition of a variable is intended to indicate that there is an *indefinitely large* number of distinct such. Human limitations being what they are, we have room and time to list only a finite number; so we add ‘. . .’. Since in practice we rarely need more than four distinct variables, there is no need to specify how the list would continue. But it is well to remember that the number of variables has no *theoretical* upper limit, that if we ever need a new one we are entitled to construct it (say by adding dashes and introducing ‘ P' ’, ‘ R''' ’, etc., into our list).

Fourth, I define a *symbol (of the propositional calculus)* as *either a bracket or a logical connective or a propositional variable.* Hence any of the above marks is a symbol.

Fifth, I define a *formula* (*of the propositional calculus*) as *any sequence of symbols*. This definition needs a little explanation; in virtue of it,

(1) ' $P((- \& Q \leftrightarrow$ '
(2) ' $(P \vee -P)$ '

are both formulae, since both are sequences of symbols: (2) for example is the sequence consisting of a left-hand bracket, followed by an occurrence of the variable ' P ', followed by the connective ' $\vee$ ', followed by the connective '—', followed by a second occurrence of the variable ' P ', followed by a right-hand bracket. But

(3) ' $\vee$ P '
& $\rightarrow$
$(\rightarrow$ R
)

is *not* a formula, since it is not a *sequence* of symbols, but rather a jumble of them. A sequence requires order, which (1) and (2) possess but (3) lacks. Our normal convention for writing sequences of symbols is that they shall appear, not spaced too far apart, in the order from left to right. This is a contemporary European convention, which the reader will be relieved to see I am following in this book.

Of the whole class of formulae, some, like (1) above, might be loosely called meaningless or gibberish, while others, like (2), make sense and can be understood. It is only, of course, the second group that we want to use in our formal work, so that we must single them out, if we can, by a precise definition. Out of the totality of formulae, therefore, we define the sub-class of *well-formed formulae*, by a somewhat complex definition, which has seven clauses. To save space, we abbreviate ' well-formed formula ' to ' wff ' (plural ' wffs '), both here and hereafter.

(*a*) any propositional variable is a wff;
(*b*) any wff preceded by ' — ' is a wff;
(*c*) any wff followed by ' $\rightarrow$ ' followed by any wff, the whole enclosed in brackets, is a wff;

(*d*) like (*c*), with ' & ' replacing ' $\twoheadrightarrow$ ';

(*e*) like (*c*), with ' v ' replacing ' $\twoheadrightarrow$ ';

(*f*) like (*c*), with ' $\twoheadleftrightarrow$ ' replacing ' $\twoheadrightarrow$ ';

(*g*) if a formula is not a wff in virtue of clauses (*a*)–(*f*), then it is not a wff.

The best way to see that these clauses do successfully define a wff is to consider examples. We show that

(4) ' $(((P \twoheadrightarrow Q) \lor -Q) \twoheadleftrightarrow (--P \, \& \, Q))$ '

is a wff, as we wish it to be, in view of the definition. First, in virtue of clause (*a*),

$$' P ', ' Q '$$

are wffs, since by (*a*) all variables are wffs. By (*b*), the result of prefixing ' — ' to a wff gives a wff: hence

$$' -P ', ' -Q '$$

are wffs. But if ' $-P$ ' is a wff, as we have shown it to be, then by clause (*b*) again

$$' --P '$$

is a wff. (We could go on applying (*b*) to show that ' $---P$ ', ' $----P$ ', etc., were all wffs.) Now since ' $--P$ ' and ' Q ' have been shown to be wffs, by clause (*d*) the result of placing ' & ' between them and enclosing the whole in brackets yields a further wff: hence

(5) ' $(--P \, \& \, Q)$ '

is a wff. Again, by (*c*), since ' P ' and ' Q ' are wffs, so is

$$' (P \twoheadrightarrow Q) '.$$

Using (*e*), given that ' $(P \twoheadrightarrow Q)$ ' and ' $-Q$ ' are wffs, we have that

(6) ' $((P \twoheadrightarrow Q) \lor -Q)$ '

is a wff. Finally, using (*f*), given that (6) and (5) are wffs, we see that (4) itself is a wff: for (4) results from writing (6), followed by ' $\twoheadleftrightarrow$ ', followed by (5), the whole enclosed in brackets. Our definition has enabled us to show, step by step beginning from the smallest parts (the variables), that a complex formula such as (6) is well-formed. A careful study of the example should make clear how the technique can be generally applied.

On the other hand, it is obvious (though not too easy to prove) that no such applications of clauses (*a*) to (*f*) could ever show that (1) above—the example of 'gibberish'—is a wff. Hence, by clause (*g*), the ruling-out or *extremal* clause of the definition, (1) is *not* a wff. The force of clauses (*a*)–(*g*), taken together, is to divide the totality of formulae into two camps: those that can be obtained by applications of clauses (*a*)–(*f*), which are wffs by the definition, and those that cannot be so obtained, which by (*g*) of the definition are not wffs.

An important aspect of this definition is the insistence, in clauses (*c*)–(*f*), on introducing surrounding *brackets*. This is necessary because of ambiguities that would result from their omission. For example, we do not wish to admit as well-formed the formula '*P & Q* ⇥ *R*', because as it stands this might mean either '(*P &* (*Q* ⇥ *R*))' (expressing a conjunction with a conditional second conjunct) or '((*P & Q*) ⇥ *R*)' (expressing a conditional with a conjunction for antecedent). Our emphasis on bracket-insertion removes risks of this kind. (On the other hand, we need no such insertion of brackets in clause (*b*), and the student may profitably speculate as to why not.)

In some ways, however, the bracketing conventions imposed by the definition of a wff, though theoretically correct, are in practice a nuisance. In fact, as a result of them the vast majority of formulae exhibited in Chapter 1 are unfortunately not well-formed. They lacked the requisite outer pair of brackets. We accepted there, for example, ' −*P* ⇥ *Q* ', whilst by clause (*c*) we require ' (−*P* ⇥ *Q*) '. But our instinct was sound, if our precision was faulty: human beings cannot stand very much proliferation of brackets. A natural *practical* convention to adopt is to permit the dropping of outermost brackets, since evidently no ambiguity can result. And there is another useful practical way in which we can cut down brackets safely, as follows.

Let us *rank* the connectives in a certain order: let us agree that ' − ' 'ties more closely' than ' & ' or ' v ', that ' & ' and ' v ' 'tie more closely' than ' ⇥ ', and that ' ⇥ ' 'ties more closely' than ' ⟷ '. Thus we can safely write in practice '*P & Q* ⇥ *R*' for

'$((P \& Q) \rightarrow R)$', dropping the outer brackets by our previous convention, and dropping the inner ones by our present one: ' $\&$ ', tying more closely than '$\rightarrow$', steals the ' Q ' in ' $P \& Q \rightarrow R$ ' for a second conjunct, rather than leaving it as the antecedent of ' $Q \rightarrow R$ '. If we require the latter interpretation, we need to write ' $P \& (Q \rightarrow R)$ '. Using these conventions, we can write (4) unambiguously in the less bracket-infested form

(7) ' $(P \rightarrow Q) \vee -Q \leftrightarrow --P \& Q$ ',

where only one pair of brackets is required. (In this connection, the student should notice the difference between $--(P \& Q)$, the double negation of the conjunction of P and Q, and $--P \& Q$, the conjunction of the double negation of P and Q.)

These conventions will be adopted from now on. But it must be stressed that they are practical guides to the eye, not theoretical devices. In theory, a wff remains as defined above, complete with its outer brackets and inner pairs of the same.

So far, we have described the basic syntax of the propositional calculus: the definition given of a wff can be read as an exact account of what is to be understood by the hitherto vague notion of a sentence in the symbolism; and clauses (a)–(f) of that definition can be read as giving what are often described as the *formation rules* of the propositional calculus—the rules, that is, determining what is a properly formed expression of the language.

But there are other syntactical notions which will be important later and which it is useful to define now. The first of these is that of the *scope* of a connective. Roughly speaking, the scope of a connective in a certain formula is the formulae *linked* by the connective, together with the connective itself and the (theoretically) encircling brackets. For example, the scope of ' $\&$ ' in (4) is the wff '$(--P \& Q)$' and the scope of ' $\leftrightarrow$ ' in (4) is the wff (4) itself: in general, the scope of any connective is a wff. More strictly, we need to define the scope of an *occurrence* of a connective in a certain wff: in (4) there are three occurrences of ' $-$ '; the scope of the first occurrence (reading from left to right) is ' $-Q$ ', the scope of the second is ' $--P$ ', and the scope of the third is ' $-P$ '. The scope is what a particular occurrence of a connective controls. A precise definition of scope is as follows: the *scope* of an occurrence of a connective in a wff is *the shortest wff in which that occurrence appears*.

Consider, for example, the (sole occurrence of) ' v ' in (4): this appears, within (4), in such formulae as:

(i) ') v — '
(ii) ' $\rightarrow Q$) v — Q) '
(iii) ' (($P \rightarrow Q$) v — Q) '
(iv) ' ((($P \rightarrow Q$) v — Q) $\leftrightarrow$ (— — '.

The shortest formula in which it appears which is also a *well-formed* formula by clauses (a)–(f) above is (iii), and this is in fact the scope of that occurrence of ' v '. Even if the definition of scope seems a bit queer, the intuitive content of the notion should be obvious.

In terms of scope, we may define a second important syntactical notion, that of one (occurrence of a) connective being *subordinate*, in a certain wff, to another. One (occurrence of a) connective is *subordinate* to another if *the scope of the first is contained in the scope of the second.* For example, in (4) the ' $\rightarrow$ ' is subordinate to the ' v ', and the ' v ' and the ' & ' are both subordinate to the ' $\leftrightarrow$ '. The first ' — ' is subordinate to ' v ', but not to ' $\rightarrow$ '; the second '—' is subordinate to ' & ' but not to ' v '; the third '—' is subordinate to the second ' — ', and so to ' & ' and ' $\leftrightarrow$ ', but not to ' $\rightarrow$ ' or ' v '. In any wff, there is exactly one connective to which all other connective-occurrences are subordinate, which is in fact the connective of widest scope. This is called the *main connective*, and its scope is the whole wff. For example, in (4) the main connective is ' $\leftrightarrow$ ', and in (2) and (6) it is ' v '.

When we prove, by application of clauses (b)–(f), that a certain formula is well-formed, we need to proceed from subordinate to subordinating connectives. Thus, in proving (4) to be well-formed, we establish that ' $-P$ ' is well-formed before we prove that ' $--P$ ' is; and that ' $--P$ ' is before we prove that ' ($--P$ & Q) ' is; and so on—at each step introducing a connective which subordinates or has in its scope the previously introduced connectives. From this point of view, the notions of scope and subordination as well as clauses (a)–(f) are ways of indicating the *natural structure* of a wff.

With the notion of a wff clear in our minds, we can readily define a *sequent-expression* (an expression which expresses a sequent, in the sense of the last chapter). As before, let us call ' $\vdash$ ' the *assertion-sign*, and let A_1, A_2, . . ., A_n, B be any set of wffs. Then

$$A_1, A_2, \ldots, A_n \vdash B$$

is a sequent-expression. In other words, write down any (finite) number of wffs, with commas between them; add to the right the assertion-sign, and follow this by any wff; the result is a sequent-expression. In the last chapter, at least 36 sequent-expressions are proved to express valid sequents, corresponding to the proofs numbered 1-36.

This last definition introduces a device which is extremely helpful in logic: the device of *metalogical variables*, such as ' A_1 ', ' A_n ', ' B '. (They appeared earlier, in Chapter 1, Section 4, in the statement of *Df.* ←→.) *Propositional* variables, such as ' P ', ' Q ', have as instances propositions; numerical variables in algebra, such as ' x ', ' y ', have as instances *numbers*. But metalogical variables are of service when we wish, as we do at present, to talk about *symbols themselves*, for they have as instances *symbols* or sequences of them. When I say that A_1, A_2, . . ., A_n are to be a set of wffs, this is entirely analogous to saying, in algebra, that x_1, x_2, . . ., x_n are to be a set of numbers. We may illustrate further the usefulness of metalogical variables by restating clauses (a)–(f) in a new form (these versions have exactly the sense of the earlier ones).

> (a') any propositional variable is a wff;
> (b') if A is a wff, then —A is a wff;
> (c') if A and B are wffs, then (A → B) is a wff;
> (d') if A and B are wffs, then (A & B) is a wff;
> (e') if A and B are wffs, then (A v B) is a wff;
> (f') if A and B are wffs, then (A ←→ B) is a wff.

EXERCISE

Select formulae (say from Chapter 1), and write them out as wffs. In each case, prove them to be wffs, using the definition of a wff; state the scope of each (occurrence of a) connective; state which is the main connective, and the relations of subordination which are present between the connective-occurrences.

2 THEOREMS AND DERIVED RULES

With the syntax of the propositional calculus now described, we may turn to the first question raised in the introduction to this chapter: what devices can we develop for utilizing already proved sequents to shorten the proofs for other sequents? One of the main devices will be the introduction of *theorems* into proofs, so that we begin by explaining what these are.

As was pointed out in Chapter 1, two out of the ten rules of derivation so far introduced—CP and RAA—have the property that as a result of their application in a proof the number of assumptions marked on the left falls by one. Suppose, now, that before the application of one of these rules there is only *one* assumption on the left: then as a result of this application there will be *no* assumption on the left. This possibility was countenanced in the statement of the rules; for example, in Section 5 of the last chapter, RAA was said to permit us, given a proof of $B \& -B$ from A, to derive $-A$ on the remaining assumptions (*if any*). Here is a simple example of a proof having this feature.

37 1 (1) $P \& -P$ A
     (2) $-(P \& -P)$ 1,1 RAA

At line (1), we assume the contradiction $P \& -P$ (nothing in the rule of assumptions prevents us from assuming what we will). Hence line (1) affirms that, given this contradiction, we have a contradiction. We can thus apply RAA to derive the negation of (1) *on no assumptions at all*. Consequently, at line (2) there are *no* citations on the left-hand side.

We may state the sequent proved at line (2) of 37 very simply.

37 $\vdash - (P \& -P)$

Here the assertion-sign appears with no wffs written to the left of it, corresponding to the absence of citation on the left at line (2). The conclusions of sequents which we can prove in this form we call *theorems*; thus a theorem is *the conclusion of a provable sequent in which the number of assumptions is zero*. Instead of reading the assertion-sign as 'therefore', which is the most natural reading in the case of sequents which have assumptions, in the case of sequents provable with no assumptions we may naturally read it as 'it is a theorem that . . .'. Thus 37 states that it is a theorem that it is not the case that P and not $-P$: for example, it is a theorem that it is not the case that it is raining and it is not raining.

Most theorems of interest are obtained in fact by application of CP. For example:

38 ⊢ $P \rightarrow P$ (compare sequent 29)

  1 (1) P A
      (2) $P \rightarrow P$ 1,1 CP

39 ⊢ $P \rightarrow --P$

  1 (1) P A
  1 (2) $--P$ 1 DN
      (3) $P \rightarrow --P$ 1,2 CP

40 ⊢ $--P \rightarrow P$

  1 (1) $--P$ A
  1 (2) P 1 DN
      (3) $--P \rightarrow P$ 1,2 CP

41 ⊢ $P \& Q \rightarrow P$ (compare sequent 14)

  1 (1) $P \& Q$ A
  1 (2) P 1 &E
      (3) $P \& Q \rightarrow P$ 1, 2 CP

38 and 41, when compared with 29 and 14, suggest that a theorem can be obtained from *any* sequent proved in the last chapter simply by appending to its proof one or more steps of CP. For example:

42 ⊢ $(P \rightarrow Q) \rightarrow (-Q \rightarrow -P)$

  1   (1) $P \rightarrow Q$ A
  2   (2) $-Q$ A
  1,2   (3) $-P$ 1,2 MTT
  1   (4) $-Q \rightarrow -P$ 2,3 CP
        (5) $(P \rightarrow Q) \rightarrow (-Q \rightarrow -P)$ 1,4 CP

Here, lines (1)–(4) are identical with the proof of sequent 9, $P \rightarrow Q \vdash -Q \rightarrow -P$, and the step of CP at line (5) completes the proof of 42. Similarly, three steps of CP added to the proof of sequent 4 yields:

43 $\vdash (P \rightarrow (Q \rightarrow R)) \rightarrow ((P \rightarrow Q) \rightarrow (P \rightarrow R))$

The importance of theorems resides in the fact that, since they are provable as conclusions from *no* assumptions, they are propositions which are *true simply on logical grounds*. Such truths, often called *logical truths* or *logical laws*, occupy an important place not only in logic but in philosophy also. Many of them have received special names. For example, 37 is called the *law of non-contradiction*; 38 is called the *law of identity*; 39 and 40 are sometimes called the *laws of double negation*. As an example of 38, we may consider the proposition that if it is raining then it is raining; this is true on purely logical grounds, quite independently of the actual state of the weather.

Theorems, such as $P \rightarrow P$, should be contrasted with the corresponding valid sequents with assumptions, such as $P \vdash P$. Whilst the latter are argument-frames, patterns of valid argument, the former are (logically) *true propositions*. ' It is raining; therefore it is raining ' expresses an argument, of which we can ask: is it *valid* or not? ' If it is raining, then it is raining ' expresses a proposition, of which we can ask: is it *true* or not? To confuse arguments with propositions is analogous to confusing validity with truth—a confusion I tried to eliminate in the first section of this book.

There is one further theorem of importance, which cannot be proved by a final step of CP since it is not conditional in form, called the *law of excluded middle*:

44 $\vdash P \vee -P$

1	(1) $-(P \vee -P)$	A
2	(2) P	A
2	(3) $P \vee -P$	2 vI
1,2	(4) $(P \vee -P) \& -(P \vee -P)$	3,1 &I
1	(5) $-P$	2,4 RAA
1	(6) $P \vee -P$	5 vI
1	(7) $(P \vee -P) \& -(P \vee -P)$	6,1 &I
	(8) $--(P \vee -P)$	1,7 RAA
	(9) $P \vee -P$	8 DN

We assume at line (1) the negation of the desired theorem, and aim for a contradiction. By assuming P (line (2)), we obtain a contradiction (line (4)) resting on both (1) and (2), so that (1) leads (line (5)) to $-P$. This leads to the same contradiction (line (7)), which now, however, rests solely on (1). Hence, using DN, we obtain the desired result. It is worth noting that at line (6) we find $P \lor -P$ resting *on its own negation* as assumption—given that it is not the case, it is the case; this should throw some light on the 'surprising' result 23 of the last chapter.

The law of excluded middle effectively affirms that, for any proposition, either it or its negation is the case, which is fairly evidently a logical truth. It is closely related to the law that every proposition is either true or false, and from this law it receives its name—a third or middle value between truth and falsity is excluded for all propositions. As a matter of logic, either it is raining or it is not raining: there is no third possibility. To be quite fair, it should be said that it can be and has been doubted whether this law has universal application: for example, is it true that either you have stopped beating your wife or you have not?

The proof just given is a proof of the theorem $P \lor -P$. Suppose, however, that we wished to prove $Q \lor -Q$; a moment's thought should convince us that, if we systematically changed each occurrence of 'P' in the given proof to 'Q', the result would be an equally sound proof of this further theorem. Suppose, again, that we wished to prove $(Q \rightarrow R) \lor -(Q \rightarrow R)$; slightly more thought should convince us that a similar change of 'P' to '$(Q \rightarrow R)$' throughout the proof will do the job. Consideration of such cases suggests that, in proving a theorem, we are implicitly proving a wide variety of other theorems closely related to the proved theorem by substitutions of the kind just instanced: so that it would be wasteful to prove these other theorems separately—it would involve virtual reduplication of the discovered proof. This in turn suggests a short cut to new results.

The matter can be made more precise by defining a *substitution-instance* of a given wff, as follows. A substitution-instance of a given wff is *a wff which results from the given wff by replacing one or more of the variables occurring in the wff throughout by some other wffs*, it being understood that each variable so replaced is replaced by the *same* wff. For example, '$(Q \rightarrow R) \lor -(Q \rightarrow R)$' is, by this definition, a substitution-instance of '$P \lor -P$', because it results from the latter wff by replacing the variable 'P' occurring in

'P v $-P$' throughout by the same wff '$(Q \rightarrow R)$'. Similarly, 'Q v $-Q$' is a substitution-instance of 'P v $-P$', and, in this case, conversely too.

Here is a more complex case. Consider the wffs:

(1) $P \rightarrow Q$ v $-(-P \& Q)$;

(2) R v $S \rightarrow P$ v $-(-(R$ v $S) \& P)$.

Then (2) is a substitution-instance of (1), because (2) results from (1) by replacing the variable 'P' at its two occurrences in (1) by '$(R$ v $S)$' and the variable 'Q' at its two occurrences in (1) by 'P'.

It is worth stressing two features of substitution which are easily and often forgotten. First, the substitution must be made *uniformly*—i.e. throughout—for each substituted variable: the *same* wff must be substituted for *every* occurrence of a given variable for a substitution-instance to result. Second, it is only on *propositional variables* that this substitution can be performed, and *not*, for example, on negated variables. Thus

(3) $-S \rightarrow Q$ v $-(S \& Q)$

is not a substitution-instance of (1), by our definition, though

(4) $-S \rightarrow Q$ v $-(--S \& Q)$

is a substitution-instance of (1): if we replace 'P' in (1) by '$-S$' throughout, we obtain (4) but not (3). Hence a substitution-instance of a wff will always be at least as long as the given wff, and none of the connectives in the given wff disappear in the substitution-instance. In an obvious though vague sense, a substitution-instance has the same broad structure as the original.

Now we can say that a proof of a theorem constitutes implicit proof of all the (indefinitely many) possible substitution-instances of that theorem. The proof of $P \rightarrow P$ (38 above) is implicitly a proof of any theorem of the form $A \rightarrow A$, for *any* wff A, and so implicitly a proof of $(-P \rightarrow Q) \rightarrow (-P \rightarrow Q)$, R v $S \rightarrow R$ v S, and so on. More precisely, suppose that the wff A expresses a theorem for which we have a proof, and suppose that B is some variable occurring in A. Then, if we systematically replace B throughout the proof of A by some other wff C, we obtain a new proof of that substitution-instance of A which results from replacing B throughout A by C. And this can be extended readily to substitution for more than one variable in A. That the new proof really is a proof—that all the applications of the rules of derivation remain correct applications after the substitution has been performed—can be seen by inspecting the rules themselves; for the rules concern only the broad structure of the wffs involved, and this structure is unaffected by substitution. We may summarize our result in the following form:

> (S1) A proof can be found for any substitution-instance of a proved theorem.

This result for theorems can be extended to sequents in general. We may define a *substitution-instance of a sequent-expression* as *any sequent-expression which results from the given sequent-expression by replacing one or more of the variables occurring in some wff in the sequent-expression throughout the sequent-expression by some other wffs*, it being understood that each variable so replaced is replaced by the *same* wff. (This definition virtually becomes the earlier definition in the limiting case that the sequent-expression contains just one wff.) For example, sequent 2 is a substitution-instance of sequent 1, and

$$(5) \quad P \to (Q \mathbin{\&} R \to -S), P, --S \vdash -(Q \mathbin{\&} R)$$

is a substitution-instance of

$$(6) \quad P \to (Q \to R), P, -R \vdash -Q,$$

obtained by substituting throughout ' $(Q \mathbin{\&} R)$ ' for ' Q ' and ' $-S$ ' for ' R '. We proved that (6) expresses a valid sequent as proof 6. We can now see that the proof of 6 constitutes implicit proof of the sequent-expression (5) also. By entirely similar reasoning, we obtain a generalization of the principle (S1):

> (S2) A proof can be found for any substitution-instance of a proved sequent.

The proof is indeed obtained by performing the relevant substitutions systematically throughout the given proof, whereupon all applications of rules of derivation remain correct applications in the new proof.

The principles (S1) and (S2) reveal an important property of our proved results, that of *generality*. We introduced symbols ' P ', ' Q ', ' R ', etc., at the outset as stand-ins for *particular* sentences of ordinary speech, which had the merit that they helped to reveal the logical form of complex sentences—a form that was shared by other sentences. We can now see that they in fact deserve the label ' variable ', since a theorem or sequent proved for P is implicitly proved for *any* proposition of the propositional calculus, just as a result in algebra containing ' x ' is implicitly a result about any number. In this way, our results, though stated for particular propositions, implicitly concern any proposition expressible in our notation, and are quite general in content.

We may take advantage of theorems and their substitution-instances to shorten proofs by the *rule of theorem introduction* (TI). This rule permits us to introduce, at any stage of a proof, a theorem already proved or a substitution-instance of such a theorem. At the right, we cite TI (or TI(S), if a substitution-instance is involved) together with the number of the theorem proved. On the left, of course, no numbers appear, since theorems depend on no assumptions. For example:

45 $P \vdash (P \& Q) \vee (P \& -Q)$

1	(1) P	A
	(2) $Q \vee -Q$	TI(S) 44
3	(3) Q	A
1,3	(4) $P \& Q$	1,3 &I
1,3	(5) $(P \& Q) \vee (P \& -Q)$	4 vI
6	(6) $-Q$	A
1,6	(7) $P \& -Q$	1,6 &I
1,6	(8) $(P \& Q) \vee (P \& -Q)$	7 vI
1	(9) $(P \& Q) \vee (P \& -Q)$	2,3,5,6,8 vE

After assuming P, we introduce (line (2)) the law of excluded middle, 44, under a substitution-instance, and then proceed by vE, assuming each disjunct of the law in turn (lines (3) and (6)), and obtaining the desired conclusion from each (lines (5) and (8)). When we apply vE at line (9), the conclusion rests only on P, since the disjunction at (2), being a theorem, rests on no assumptions.

46 $P \rightarrow Q \vdash P \& Q \leftrightarrow P$

1	(1) $P \rightarrow Q$	A
	(2) $P \& Q \rightarrow P$	TI 41

3	(3) P	A
1,3	(4) Q	1,3 MPP
1,3	(5) $P \& Q$	3,4 &I
1	(6) $P \rightarrow P \& Q$	3,5 CP
1	(7) $(P \& Q \rightarrow P) \& (P \rightarrow P \& Q)$	2,6 &I
	(8) $P \& Q \leftrightarrow P$	7 $Df. \leftrightarrow$

To obtain the biconditional $P \& Q \leftrightarrow P$, we aim separately at the two conditionals $P \& Q \rightarrow P$ and $P \rightarrow P \& Q$; but the first is a proved theorem, 41, which we therefore introduce directly by TI. Conjoining 27 and 46, we have the interderivability result

47 $P \& Q \leftrightarrow P \dashv\vdash P \rightarrow Q$.

The rule TI is *not* a new fundamental rule of derivation: it does not enable us to prove sequents which we cannot otherwise prove by applications of our basic ten rules; it merely enables us to prove *more briefly* further results by using results already proved. In the case of 45, for example, we could prefix the proof given by 8 lines, corresponding to the first 8 lines of the proof of 44 but with ' Q ' in place of ' P ', and then continue as before, renumbering (1) to (9) as (9) to (17). In place of TI(S) 44, we would read on the right 8 DN (compare line (9) of 44), and thus obtain a complete, if lengthy, proof of 45 from our basic rules. Whenever a theorem is introduced by TI, we can prefix the proof given by a proof of the theorem from basic rules, and thus transform the proof into a lengthier proof from first principles: only a certain renumbering of lines is involved. Rules of this character, which expedite our proof-techniques but can be shown not to increase our derivational power, are called *derived rules*, in contrast to our basic ten rules, which may be called *primitive rules*.

Having seen this use of theorems to shorten proofs, we naturally ask whether an analogous rule will enable us to use sequents already proved. For example, suppose that we have proved, on certain assumptions, $P \rightarrow Q$. Then, by sequent 9 ($P \rightarrow Q \vdash -Q \rightarrow -P$), we should be able to conclude, without special proof, $-Q \rightarrow -P$ on the same assumptions. Or suppose that we have proved, on various assumptions, $P \rightarrow Q$, $Q \rightarrow R$, and P. Then, by sequent 3 ($P \rightarrow Q$, $Q \rightarrow R$, $P \vdash R$), we should be able to conclude, without special proof, R on the pool of these assumptions. And this should apply not only to the sequents actually proved but to any substitution-instances of them too, in virtue of (S2).

The *rule of sequent introduction* (SI), again a derived not a primitive rule, enables us to do just this. It is a little complex both to state and to justify in full generality, but its main function should be clear from examples. Suppose that we have as conclusions in a proof $A_1, A_2, \ldots, A_n$, on various assumptions, and suppose that $A_1, A_2, \ldots, A_n \vdash B$ is a (substitution-instance of a) sequent for which we already have a proof; then SI permits us to draw B as a conclusion on the pool of the assumptions on which $A_1, A_2, \ldots, A_n$ rest. SI may be justified as follows. By hypothesis (and (S2) if necessary), we have a proof using only primitive rules of

(i) $A_1, A_2, \ldots, A_n \vdash B$.

Hence, by n successive steps of CP added to the proof, we can prove as a theorem

(ii) $A_1 \rightarrow (A_2 \rightarrow (\ldots (A_n \rightarrow B) \ldots))$

(the conditional theorem corresponding to the sequent in the way in which the conclusion of 43 above corresponds to 4). Hence by TI we can introduce (ii) into the proof given with conclusions A_1, $A_2, \ldots, A_n$, as a new line resting on no assumptions. Now, by n successive steps of MPP, using in turn $A_1, A_2, \ldots, A_n$ as antecedents of given conditionals, we can draw as conclusion B. Evidently B will depend, as assumptions, on any propositions on which any of $A_1, A_2, \ldots, A_n$ depends. This justifies SI, in the sense that it shows how any proof using SI can be systematically transformed into a proof of the same sequent using only primitive rules—the step of TI involved can, as we already know, be eliminated in favour of these rules.

48 $-P \vee Q \vdash P \rightarrow Q$

1	(1)	$-P \vee Q$	A
1	(2)	$-(--P \ \& -Q)$	1 SI(S) 36(*a*)
1	(3)	$--P \rightarrow Q$	2 SI(S) 35(*b*)
4	(4)	P	A
4	(5)	$--P$	4 DN

| 1,4 | (6) Q | 3,5 MPP |
| 1 | (7) $P \rightarrow Q$ | 4,6 CP |

A substitution-instance of 36(*a*) is $-P \vee Q \vdash -(--P \& -Q)$, and a substitution-instance of 35(*b*) is $-(--P \& -Q) \vdash --P \rightarrow Q$: these two sequents are used to obtain (2) from (1) and (3) from (2) by SI. The rest of the proof is then immediate. Together with Exercise 1.5.1(*i*), 48 yields

49 $P \rightarrow Q \dashv\vdash -P \vee Q$.

50 $P \vdash Q \rightarrow P$

1	(1) P	A
1	(2) $-Q \vee P$	1 vI
1	(3) $Q \rightarrow P$	2 SI(S) 48

51 $-P \vdash P \rightarrow Q$

1	(1) $-P$	A
1	(2) $-P \vee Q$	1 vI
1	(3) $P \rightarrow Q$	2 SI 48

52 $-P, P \vee Q \vdash Q$

1	(1) $-P$	A
2	(2) $P \vee Q$	A
3	(3) P	A
1	(4) $P \rightarrow Q$	1 SI 51
1,3	(5) Q	3,4 MPP
6	(6) Q	A
1,2	(7) Q	2,3,5,6,6 vE

53 $-Q, P \vee Q \vdash P$

(Proof similar to 52.)

54 $\vdash (P \rightarrow Q) \vee (Q \rightarrow P)$

	(1) $P \vee -P$	TI 44
2	(2) P	A
2	(3) $Q \rightarrow P$	2 SI 50

59

2	(4) $(P \rightarrow Q) \vee (Q \rightarrow P)$	3 vI
5	(5) $-P$	A
5	(6) $P \rightarrow Q$	5 SI 51
5	(7) $(P \rightarrow Q) \vee (Q \rightarrow P)$	6 vI
	(8) $(P \rightarrow Q) \vee (Q \rightarrow P)$	1,2,4,5,7 vE

An interesting feature of this last series of results is its progressive nature: once 48 was proved, it was used to obtain 50 and 51, which in turn were employed in the proofs of 52 and 54. It should be clear by now that TI and SI are powerful devices for generating new theorems and sequents out of old. Our work now has the progressive character of a mathematical theory such as Euclidean geometry.

Of the latest results, 50 and 51 are sometimes called *the paradoxes of material implication*. To see their paradoxical flavour, bear in mind that Q in 50 and 51 may be *any* proposition, even one quite unrelated in content to P. Thus 50 enables us to conclude from the fact that Napoleon was French that if the moon is blue then Napoleon was French; and 51 enables us to conclude from the fact that Napoleon was not Chinese that if Napoleon was Chinese then the moon is blue. The name ' material implication ' was given by Bertrand Russell to the relation between P and Q expressed in our symbolism by '$P \rightarrow Q$'; we have been reading this ' if P then Q ', but it is clear from 50 and 51 that ' $\rightarrow$ ' has logical properties which we should not ordinarily associate with ' if . . . then . . .'. This discrepancy is chiefly brought about by the fact that, before we would ordinarily accept ' if P then Q ' as true, we should require that P and Q be connected in thought or content, whilst, as 50 and 51 show, no such requirement is imposed on the acceptance of ' $P \rightarrow Q$ '. However, whilst admitting that this discrepancy exists, we may continue safely to adopt '$P \rightarrow Q$' as a rendering of 'if P then Q ' *serviceable for reasoning purposes*, since, as will emerge in Section 4, our rules at least have the property that they will never lead us from true assumptions to a false conclusion. And any reader who is inclined not to accept the validity of 50 and 51 is asked either to suspend judgement until this fact has been established or to indicate exactly which step in their proof he regards as faulty and which rule of derivation he thinks is unsafe and why. (A natural reply is that the step of vI at line (2) of each proof is unsound; but compare the justification of vI in Chapter 1, Section 3. Anyway, 50 and 51 can be proved using only the rules A, &I, &E, RAA, DN, and CP, in each case in nine lines; it is an instructive exercise to discover these ' independent ' proofs, since they reveal how difficult it is to ' escape ' the paradoxes.) Along with 23, therefore, we may classify 50 and 51 as some of the more surprising consequences of our primitive rules. 54 is a less well-known paradox: it claims as a logical truth that, for any propositions *P* and *Q*, it is either the case that if *P* then *Q* or the case that if *Q* then *P*. Either if it is raining it is snowing or if it is snowing it is raining.

The principle of reasoning associated with 52 and 53 has the medieval name *modus tollendo ponens*. This is the fourth medieval *modus* I have mentioned, and the last there is, so this is a good place to bring them together.

(i) *Modus ponendo ponens* is the principle that, if a conditional holds and also its antecedent, then its consequent holds;

(ii) *Modus tollendo tollens* is the principle that, if a conditional holds and also the negation of its consequent, then the negation of its antecedent holds;

(iii) *Modus ponendo tollens* is the principle that, if the negation of a conjunction holds and also one of its conjuncts, then the negation of its other conjunct holds;

(iv) *Modus tollendo ponens* is the principle that, if a disjunction holds and also the negation of one of its disjuncts, then the other disjunct holds.

(i) and (ii) have been embodied in our primitive rules MPP and MTT. Clearly, in virtue of SI and 52 and 53, a rule analogous to *modus tollendo ponens*, which we may call MTP, can be framed; this, as a *derived* rule, will merely be a special case of SI. It runs: given a disjunction and the negation of one disjunct, then we are permitted to derive the other disjunct as conclusion. When required, this rule will in fact be cited as MTP. Similarly, in virtue of SI and 34 and the readily proved $Q, -(P \& Q) \vdash -P$, we may formulate, as a special derived rule, MPT: given the negation of a conjunction and one of its conjuncts, then we are permitted to derive the negation of the other conjunct as conclusion.

In connection with the *modi*, it is finally worth noting that MTT need not have been taken as a primitive rule, but can be obtained as a derived rule from the others. Thus:

55 $P \rightarrow Q, -Q \vdash -P$

1	(1) $P \rightarrow Q$	A
2	(2) $-Q$	A
3	(3) P	A
1,3	(4) Q	1,3 MPP
1,2,3	(5) $Q \,\&\, -Q$	2,4 &I
1,2	(6) $-P$	3,5 RAA

We prove 55 without using MTT. In view of SI, 55 can be used to give exactly the effect of MTT as a derived rule. This would be of interest if we were trying to reduce our primitive rules to as small a number as possible—an important consideration in certain areas of logic.

Apart from the special cases of MTP and MPT, the most rewarding sequents for use with SI are the various forms of *de Morgan's laws*, as they are called, namely 36 and Exercise 1.5.1(*f*)–(*h*), which enable us to transform negated conjunctions and disjunctions into non-negated disjunctions and conjunctions respectively. Also worth remembering are 49 (enabling us to change conditionals into disjunctions), 35 (enabling us to change conditionals into negated conjunctions), Exercise 1.5.1(*e*) (enabling us to change conjunctions into negated conditionals), and Exercise 1.5.1(*c*) and (*d*) (the so-called *distributive laws*). Often it helps to introduce 44 and proceed by vE, as in the proof of 54. And the trick of using the paradoxes 50 and 51, as in the same proof, should be borne in mind.

EXERCISES

1   Using only the 10 primitive rules, prove the following sequents:

(*a*) $\vdash (Q \rightarrow R) \rightarrow ((P \rightarrow Q) \rightarrow (P \rightarrow R))$

(*b*) $\vdash P \rightarrow (Q \rightarrow P \,\&\, Q)$

(*c*) $\vdash (P \rightarrow R) \rightarrow ((Q \rightarrow R) \rightarrow (P \lor Q \rightarrow R))$

(*d*) $\vdash (P \rightarrow Q \,\&\, -Q) \rightarrow -P$

(*e*) $\vdash (-P \rightarrow P) \rightarrow P$

2 The following are valid sequents, because they are substitution-instances of sequents already proved in this book. For each, cite by number the proved sequent of which it is a substitution-instance, and what substitutions have been used:

(a) $(P \rightarrow Q) \rightarrow P, P \rightarrow Q \vdash P$

(b) $--P \rightarrow ---P, ----P \vdash -P$

(c) $-P \& (Q \& R) \rightarrow Q \vee P \vdash -P \rightarrow (Q \& R \rightarrow Q \vee P)$

(d) $(-Q \rightarrow Q) \rightarrow -(-Q \rightarrow Q) \vdash -(-Q \rightarrow Q)$

(e) $-(S \vee P) \vdash S \vee P \rightarrow (P \& Q \leftrightarrow R \vee -S)$

3 Prove by primitive rules alone:

(a) $\vdash P \rightarrow P \vee Q$

Using this result, prove by primitive rules and TI:

(b) $Q \rightarrow P \vdash P \vee Q \leftrightarrow P$

In view of Exercise 1.4.1(e) this gives:

(c) $P \vee Q \leftrightarrow P \dashv\vdash Q \rightarrow P$

4 Prove, using primitive rules and SI in connection with 50:

(a) $P \& Q \dashv\vdash P \& (P \leftrightarrow Q)$

5 Using primitive or derived rules, together with any sequents or theorems already proved, prove:

(a) $\vdash P \vee (P \rightarrow Q)$

(b) $\vdash (P \rightarrow Q) \vee (Q \rightarrow R)$

(c) $\vdash ((P \rightarrow Q) \rightarrow P) \rightarrow P$

(d) $-Q \vdash P \rightarrow (Q \rightarrow R)$

(e) $P, -P \vdash Q$

(f) $P \vee Q \dashv\vdash -P \rightarrow Q$ (cf. Ex. 1.5.1(j))

(g) $-(P \rightarrow Q) \dashv\vdash P \& -Q$

(h) $(P \rightarrow Q) \rightarrow Q \dashv\vdash P \vee Q$

(i) $(P \rightarrow Q) \vee (P \rightarrow R) \dashv\vdash P \rightarrow Q \vee R$

(j) $P \rightarrow Q \dashv\vdash (P \leftrightarrow Q) \vee Q$

(k) $Q \vdash P \& Q \leftrightarrow P$

(l) $-Q \vdash P \vee Q \leftrightarrow P$

6 Let A and B be any propositions expressible in the propositional calculus notation.

   (i) Show that $A \vdash B$ is provable by our rules if and only if it is provable that $\vdash A \rightarrow B$;

(ii) Show that A ⊣⊢ B is provable if and only if it is provable that ⊢ A ⟷ B.

7   In effect, our rule DN is two rules combined: (i) from A to derive −−A, and (ii) from −−A to derive A. Show that (i) can be obtained as a derived rule from the other primitive rules (compare the corresponding demonstration for the rule MTT, sequent 55).

3 TRUTH-TABLES

The last section has answered the first question raised at the beginning of this chapter.   To help answer the other two questions (Are our rules of derivation safe? Are they complete?), we approach in this section the propositional calculus in a quite new way, by the technique of *truth-tables*. This technique will also incidentally afford us a method of showing the *invalidity* of sequents, whereas the rules of derivation merely show their validity. Truth-tables are easy to master, so our treatment here will be brisk.

The truth-table method is a method for *evaluating* wffs: we assign values (called *truth-values*) to the variables of a wff, and proceed by means of given tables to calculate the value of the whole wff. We may usefully compare the corresponding mathematical procedure for evaluating algebraic expressions, say

(1) $(x + y)z - (y + z)(y + x).$

Let us assign the value 10 to x, 3 to y, and 5 to z.   By substitution, we obtain

(2) $(10 + 3)5 - (3 + 5)(3 + 10).$

Computation by given tables yields successively

(3) $13 \times 5 - 8 \times 13;$

(4) $65 \quad - \quad 104;$

(5) $- 39.$

The result at (5) is the value of the whole expression (1) for the assignment of values to the variables $x = 10$, $y = 3$, $z = 5$.

In the case of wffs of the propositional calculus, there are only two possible values which variables are permitted to take, the *true* and the *false*, which we mark by ' T ' and ' F ' respectively. Our assumption that there are only these two possibilities is in effect the assumption that every proposition is either true or false, and corresponds to the law of excluded middle (theorem 44). These two values are called *truth-values*.

In order to compute the truth-value of a whole wff for a given assignment of truth-values to its variables, we need tables (called *matrices*) for each logical connective showing how the value of a complex formula is determined by the values of its parts; the matrices correspond to multiplication- and addition-tables, with the difference that, since there are only two values to consider, they can be given *in toto* whilst the mathematical tables can only be given in part (we learn, e.g., multiplication-tables *up to* 12). The matrices are:

A	−A
T	F
F	T

	A → B	T F
A	T	T F
	F	T T

	A & B	T F
A	T	T F
	F	F F

	A v B	T F
A	T	T T
	F	T F

	A ←→ B	T F
A	T	T F
	F	F T

The matrix for ' — ' is motivated by the consideration that the negation of a true proposition is false and the negation of a false proposition is true. The matrix for ' & ' is motivated by the consideration that a conjunction A & B is true if both A and B are true, but otherwise false. The matrix for ' v ' is motivated by the consideration that a disjunction A v B is true if at least one of A and B is true, but otherwise false. The matrices for ' → ' and ' ←→ ' have peculiarities, and we shall consider their motivation a little later.

To illustrate the use of these matrices, let us evaluate the wff

(6) $(P \rightarrow Q) \vee -Q \leftrightarrow --P \& Q$

for the assignment of values $P = T$, $Q = F$. Substitution gives

(7) $(T \rightarrow F) \vee -F \leftrightarrow --T \& F,$

and computation by the matrices yields successively

(8) $F \vee T \leftrightarrow -F \& F;$

(9) $T \leftrightarrow T \& F;$

(10) T $\longleftrightarrow$ F;
(11) F.

Thus the whole wff takes the value F for the given assignment. We can conveniently abbreviate the working to one line of calculation performed below the wff, thus:

$$
\begin{array}{cc|c}
P & Q & (P \rightarrow Q)\,\text{v} - Q \longleftrightarrow --P \,\&\, Q \\
\hline
T & F & T\ F\ F\ \ T\ \ TF \ \ \ \ F \ \ \ \ T\ FT\ F\ F
\end{array}
$$

On the left, we list the variables in the wff, and under them write the assignment in question. We then transcribe the value of each variable to each occurrence of it in the wff, and compute in stages the value of the wff as a whole. It is worth noting that these stages correspond to the relations of subordination among the connectives (in the sense of Section 1): we obviously need to compute the value of a subordinate connective before we can compute its subordinating connective. Finally, the value of the wff itself appears under the main connective—in this case ' $\longleftrightarrow$ '.

A *truth-table test* on a wff is an evaluation of the wff for *every possible* assignment of truth-values to its variables. If a wff has only one variable, say ' P ', then there are two possible assignments, $P = T$ and $P = F$. If a wff has two variables, say P and Q, then there are four possible assignments: (i) $P = T$, $Q = T$; (ii) $P = T$, $Q = F$; (iii) $P = F$, $Q = T$; (iv) $P = F$, $Q = F$. In general, if a wff has n variables, there will be 2^n possible assignments.

We may display a truth-table test on a wff by performing the evaluation for each assignment on a separate line under the wff. For example:

$$
\begin{array}{cc|c}
P & Q & (P \rightarrow Q\,\text{v} - Q \longleftrightarrow --\ P \,\&\, Q \\
\hline
T & T & T\ T\ T\ TF\ T \ \ \ T \ \ \ \ TF\ T\ T\ T \\
T & F & T\ F\ F\ TT\ F \ \ \ F \ \ \ \ TF\ T\ F\ F \\
F & T & F\ T\ T\ TF\ T \ \ \ F \ \ \ \ FT\ F\ F\ T \\
F & F & F\ T\ F\ TT\ F \ \ \ F \ \ \ \ FT\ F\ F\ F
\end{array}
$$

(The second line of this test is, of course, identical with the line computed earlier for the same wff.) When the test is given in this form, the values of the whole wff are shown in the column under the main connective, which we may call the *main column*.

The student ought quickly to learn from these remarks how to perform truth-table tests for himself, and he will probably find himself devising short-cuts (for example, ' $- -$ ' may be ignored wherever it occurs). It is useful to have a standard order for the different possible assignments to variables. The standard order for two variables is shown above; the standard order for three variables is shown in the following test, and analogous standard orders for assignments to four or more variables can be devised by the reader.

P	Q	R	$(P$ & $-Q)$ v $R \rightarrow (R$ & $-Q \rightarrow -P)$
T	T	T	T F FT TTT TF FTT FT
T	T	F	T F FT FFT FF FTT FT
T	F	T	T T TF TTF TT TFF FT
T	F	F	T T TF TFT FF TFT FT
F	T	T	F F FT TTT TF FTT TF
F	T	F	F F FT FFT FF FTT TF
F	F	T	F F TF TTT TT TFT TF
F	F	F	F F TF FFT FF TFT TF

For wffs containing four variables, sixteen lines are required; in the case of five variables thirty-two; and so on. Hence actually to perform a truth-table test becomes increasingly impracticable as the number of variables goes up. Despite this, it is worth mentioning that, for *any* number of variables, truth-table testing is an entirely mechanical procedure. Machines can be, and have in fact been, called upon to perform truth-table tests (they usually prove better at the job than human beings). In this respect truth-table testing contrasts with the discovery of proofs in general; a proof, *once discovered*, can mechanically be *checked* (we have indeed specified our rules of derivation partly so that this should be so). But they cannot in general be discovered by machine.[1]

In pursuance now of a motivation for the matrices for ' $\rightarrow$ ' and ' $\leftrightarrow$ ', we note that, in virtue of sequent 49, $P \rightarrow Q$ and $-P$ v Q are interderivable. We may naturally expect, therefore, that they will have the same values for the same assignments to their variables. If we truth-table test ' $-P$ v Q ', we obtain

---

[1] Actually, it turns out, in view of the results in Section 5 of this chapter, that proofs *can* be mechanically discovered at the level of the propositional calculus. But it is known that there is no mechanical procedure for proof-discovery in the predicate calculus, to which we turn in Chapters 3 and 4.

P	Q	$-P \lor Q$
T	T	F T T T
T	F	F T F F
F	T	T F T T
F	F	T F T F

which reveals the same values, assignment for assignment, as those for ' $P \rightarrow Q$ '.

In virtue of 35, $P \rightarrow Q$ and $-(P \& -Q)$ are also interderivable, and a truth-table test on ' $-(P \& -Q)$ ' gives just the same result. Intuitively, the most important thing about the proposition that if P then Q is that we certainly wish it to be false if P is true and Q false, and this at least is secured by the matrix. The other values, which admittedly seem rather arbitrary, may be justified by the interderivability results just cited. Once the matrix for ' $\rightarrow$ ' is agreed, the matrix for ' $\leftrightarrow$ ' follows, since we want $P \leftrightarrow Q$ to be equivalent to $(P \rightarrow Q) \& (Q \rightarrow P)$. Actual truth-table computation of this latter wff gives just the matrix for ' $\leftrightarrow$ ': a fact which the reader should check for himself.

On the basis of a truth-table test, we can classify all wffs of the propositional calculus in one of three ways. If a wff takes the value T for all possible assignments of truth-values to its variables, it is said to be *tautologous* or a *tautology*. If a wff takes the value F for all possible assignments of truth-values to its variables, it is said to be *inconsistent* or an *inconsistency*. If for at least one assignment it takes the value T and for at least one assignment it takes the value F, it is said to be *contingent* or a *contingency*. Clearly, every wff is one and only one of these three things, and which it is can be read off directly from the main column of the truth-table: if all T's appear there, it is tautologous; if all F's, inconsistent; if at least one T and at least one F, contingent. (Sometimes, in place of ' inconsistent ', the term ' contradictory ' or ' self-contradictory ' is used; we have avoided this here, since a contradiction was otherwise defined in Chapter 1, Section 3.) We further say that a wff is *consistent* if it is either tautologous or contingent; thus a wff is consistent if and only if for at least one assignment it takes the value T, that is, if and only if it is not inconsistent. Similarly, a wff is *non-tautologous* if it is either contingent or inconsistent.

Of these concepts by far the most important is that of a tautology.

Tautologies are true whatever the truth-values of their constituent variables, which is to say that they are true simply in virtue of their logical form, their structure in terms of logical connectives. Hence tautologies, like theorems, can be viewed as expressing logical laws, propositions true simply on logical grounds. This raises the question of the relation between theorems (defined in terms of our rules of derivation) and tautologies (defined in terms of the truth-table test). This question is answered in the following two sections, where we show that all theorems are tautologies and that all tautologies can be proved as theorems from our rules.

There are six logical relationships between two propositions which are recognized in traditional logic, and we may use in certain cases the truth-table method to establish the presence or absence of these. For each relationship I first define it, and then show how truth-tables may be applied in the relevant cases.

(i) Two propositions A and B are traditionally called *contrary* if they are never both true (though they may both be false); that is, if whenever one is true the other is false. Now to say that they are never both true is to say that their conjunction is always false, which is to say that the negation of their conjunction, −(A & B), is always true. Hence, *if A and B are expressible in propositional calculus notation*,[1] we may discover whether A and B are contrary by subjecting − (A & B) to a truth-table test: if it is tautologous, they are; if not, they are not. For to say that − (A & B) is *always* true is to say that it is true for all possible assignments of truth-values to its variables.

(ii) Two propositions A and B are *subcontrary* if they are never both false (though they may both be true); that is, if whenever one is false the other is true. Now to say that they are never both false is to say that always at least one of them is true, which is to say that their disjunction, A v B, is always true. Hence, as in (i), if A and B are expressible in propositional calculus notation, A and B are subcontrary if and only if A v B is a tautology.

(iii) A proposition A *implies* a proposition B if whenever A is true B is true (but not necessarily conversely). It is easy to see from the

[1] We impose this restriction here and in the other cases because for more complex propositions, of the kind studied in Chapters 3 and 4, propositional calculus notation is inadequate, and truth-table machinery can no longer be used to test for these relationships.

matrix for ' $\rightarrow$ ' that, if A and B are expressible in propositional calculus notation, A implies B if and only if A $\rightarrow$ B is tautologous. In this case, it is sometimes said that A is *superimplicant* or *super-alternate* to B.

(iv) A proposition A *is implied by* a proposition B if whenever B is true A is true (but not necessarily conversely). It is again easily shown that, if A and B are expressible in propositional calculus notation, A is implied by B if and only if B $\rightarrow$ A is tautologous. In this case, it is often said that A is *subimplicant* or *subalternate* to B.

(v) Two propositions A and B are *equivalent* if whenever A is true B is true and whenever B is true A is true. It is easily shown that, if A and B are expressible in propositional calculus notation, A is equivalent to B if and only if A $\leftrightarrow$ B is tautologous (from this standpoint, A $\leftrightarrow$ B affirms really that A and B have the same truth-value). In this case, A and B are sometimes said to be *coimplicant*.

(vi) Two propositions A and B are *contradictory* if they are never both true and never both false either; that is, if whenever one is true the other is false and whenever one is false the other is true. Now to say that they always *dis*agree in truth-value is to say that it is always not the case that they have the same truth-value. Hence, if A and B are expressible in propositional calculus notation, A and B are contradictory if and only if $-$(A $\leftrightarrow$ B) is tautologous.

If none of these six relationships holds between A and B, then A and B are *independent*. If A and B are expressible in propositional calculus notation, a series of truth-table tests will establish which relationships hold between them and which do not.

This piece of traditional logic can be viewed as an (inadequate) attempt to list all possible relations between two propositions which can be defined in terms of truth-values. A simple mathematical count shows that there are sixteen such relations, which we exhibit in the following table:

A	B	a	b	c	d	e	f	g	h	i	j	k	l	m	n	o	p
T	T	T	T	T	T	T	T	T	T	F	F	F	F	F	F	F	F
T	F	T	T	T	T	F	F	F	F	T	T	T	T	F	F	F	F
F	T	T	T	F	F	T	T	F	F	T	T	F	F	T	T	F	F
F	F	T	F	T	F	T	F	T	F	T	F	T	F	T	F	T	F

The sixteen columns (a)–(p) give all possible distinct matrices there can be of the kind associated with ' $\rightarrow$ ', ' & ', ' v ', and ' $\leftrightarrow$ ' earlier in this section. Thus we recognize in (e) the matrix for ' $\rightarrow$ ', in (h) the matrix for ' & ', in (b) the matrix for ' v ', and in (g) the matrix for ' $\leftrightarrow$ '. Now $P \rightarrow Q$, $P \& Q$, $P \vee Q$, $P \leftrightarrow Q$ can be described as *functions* of P and Q, in just the sense that $x + y$ is a function in algebra of x and y. To distinguish such functions from mathematical functions, we call them *truth-functions*. Then the above table lists *all possible truth-functions of two variables*.

If A and B are expressible in propositional calculus notation, the claim that A implies B is just the claim that the function of A and B defined by (e), namely A $\rightarrow$ B, is tautologous. Similarly, the claim that A and B are subcontrary is just the claim that the function of A and B defined by (b), namely A v B, is tautologous. Since the distinct functions defined in this way by columns (a)–(p) are exhaustive, there are exactly sixteen distinct such claims of relationship between A and B that can be made: namely, for each function, the claim that that function of A and B is tautologous. In fact, to say that A and B are equivalent is to say that the function defined by (g), A $\leftrightarrow$ B, is tautologous. To say that A is implied by B is to say that the function defined by (c) is tautologous; for a test of B $\rightarrow$ A for the four possible combinations of truth and falsity yields column (c), showing it to be the function defined by that column. To say that A and B are contrary is to say that the function defined by (i) is tautologous; for a test of $-$(A & B) yields column (i), showing it to be the function defined by that column. Finally, to say that A and B are contradictory is to say that the function defined by (j) is tautologous; for a test of $-$(A $\leftrightarrow$ B) yields column (j), showing it to be the function defined by that column. Thus with each traditional relationship (i)–(vi) is associated one column from (a)–(p), and there are ten further similarly definable relationships which are not traditionally distinguished.

An interesting further feature of the sixteen functions is that, for each of them, an expression can be found employing only ' $\rightarrow$ ' and ' $-$ ' which is equivalent to it. For the record, we list a possible such set of expressions:

(a) A $\rightarrow$ A	(p) $-$(A $\rightarrow$ A)
(b) $-$A $\rightarrow$ B	(o) $-$($-$A $\rightarrow$ B)

(c) $B \rightarrowtail A$
(d) A
(e) $A \rightarrowtail B$
(f) B
(g) $-((A \rightarrowtail B) \rightarrowtail -(B \rightarrowtail A))$
(h) $-(A \rightarrowtail -B)$

(n) $-(B \rightarrowtail A)$
(m) $-A$
(l) $-(A \rightarrowtail B)$
(k) $-B$
(j) $(A \rightarrowtail B) \rightarrowtail -(B \rightarrowtail A)$
(i) $A \rightarrowtail -B$

Actual testing of these expressions, by the matrices for ' $\rightarrowtail$ ' and ' $-$ ', will reveal them to be equivalent to the sixteen listed functions. In view of the fact that ' $P \rightarrowtail Q$ ' is equivalent to both ' $-P \lor Q$ ' and ' $-(P \& -Q)$ ', as we saw earlier when defending the matrix for ' $\rightarrowtail$ ', it is not surprising that, for each function, an expression can be found containing only ' $\lor$ ' and ' $-$ ' (or only ' $\&$ ' and ' $-$ ') equivalent to it.

One final application of truth-table procedures deserves mention here. It connects with substitution-instances, as defined in the previous section. We first observe that *any substitution-instance of a tautology is itself tautologous*. Although I shall not strictly prove this proposition here, it ought to be fairly obvious in any case. If A is a tautology, A takes the value T for *all possible* assignments of truth-values to its variables. In the process of substitution on A, we replace each variable *systematically* by a certain wff. When a substitution-instance of A is subjected to a truth-table test, that wff, for each assignment, will take throughout the *same* value, T or F. Hence from this point onwards the test will follow the same pattern as *some* line of the original truth-table test, and must end by assigning T to the substitution-instance as well. In a similar way, but considering F instead of T, we see that *any substitution-instance of an inconsistency is itself inconsistent*.

In the case of contingencies, however, a different situation arises. We can show that, *for any contingent wff, a substitution-instance can be found which is tautologous, and a substitution-instance can be found which is inconsistent*. I do not prove this here, but I shall indicate a quite general procedure for finding such substitution-instances, and if this procedure is followed in particular cases the reader will soon see why it works.

Let A be a wff which is contingent. Then for at least one assignment of values to its variables it takes the value T. Select one particular such assignment, and for each variable in A substitute any tautology (say '$P \rightarrow P$') if the variable takes the value T in that assignment, and any inconsistency (say '$-(P \rightarrow P)$') if the variable takes the value F in that assignment. Then the resulting substitution-instance of A will be *tautologous*. Again, since A is contingent, for at least one assignment of values to its variables it takes the value F. Select one such assignment, and then carry through *exactly* the same procedure, substituting '$P \rightarrow P$', say, for any variable with value T in the assignment and '$-(P \rightarrow P)$', say, for any variable with value F. Then the resulting substitution-instance of A will be *inconsistent*.

One simple example will suffice. '$P \vee Q$' is contingent, taking the value F if $P = $ F and $Q = $ F and otherwise the value T. Select the assignment $P = $ T, $Q = $ F, for which $P \vee Q = $ T, and accordingly substitute for 'P' '$P \rightarrow P$' and for 'Q' '$-(P \rightarrow P)$'. Then the result, '$(P \rightarrow P) \vee -(P \rightarrow P)$', is a tautologous substitution-instance of '$P \vee Q$'. Similarly, '$-(P \rightarrow P) \vee -(P \rightarrow P)$', corresponding to the assignment $P = $ F, $Q = $ F, is an inconsistent substitution-instance of the same contingent wff.

EXERCISES

1   (i) Perform truth-table tests on the following wffs, and state in each case whether the wff is tautologous, contingent, or inconsistent:

(a) $P \rightarrow P$

(b) $P \rightarrow -P$

(c) $-(P \rightarrow P)$

(d) P

(e) $-P \rightarrow (P \rightarrow Q)$

(f) $(P \leftrightarrow Q) \leftrightarrow -(P \leftrightarrow -Q)$

(g) $(P \mathbin{\&} Q) \mathbin{\&} -(P \leftrightarrow Q)$

(h) $(P \vee -Q) \mathbin{\&} -(-P \rightarrow -Q)$

(i) $(P \mathbin{\&} Q \rightarrow R) \rightarrow (P \rightarrow R) \mathbin{\&} (Q \rightarrow R)$

(j) $(P \vee Q \rightarrow R) \leftrightarrow (P \rightarrow R) \mathbin{\&} (Q \rightarrow R)$

(k) $(P \rightarrow Q) \mathbin{\&} (R \rightarrow S) \rightarrow (P \vee R \rightarrow Q \vee S)$

(ii) In the case of contingencies among (a)–(k) of (i), find a substitution-instance which is tautologous and one which is inconsistent.

2   (i) Find expressions equivalent to each of the sixteen possible truth-functions of two variables employing (a) at most '&' and '$-$', (b) at most '$\vee$' and '$-$'.

   (ii) We may denote the functions (i) and (o) in the table of possible truth-functions by ' A | B ' and ' A ⍱ B ' respectively (these are sometimes known as the ' stroke '-functions; ' A | B ' may be read ' not both A and B ' and ' A ⍱ B ' as ' neither A nor B '). Find expression equivalent to each of the sixteen possible truth-functions of two variables employing (a) only ' | ', (b) only ' ⍱ '. (hint: a test of ' $P | P$ ' and ' $P ⍱ P$ ' reveals their equivalence to ' $-P$ '. Thus ' $(P | Q)/(P | Q)$ ' is equivalent to ' $-(P | Q)$ '. and so to ' $P \& Q$ '. Similarly, ' $(P ⍱ Q) ⍱ (P ⍱ Q)$ ' is equivalent to ' $P \vee Q$ '.)

3   (i) Let A be a wff containing any number of variables but ' & ' as its sole connective: (a) show that A cannot be tautologous; (b) show that A cannot be inconsistent.

   (ii) Let A be a wff containing any number of variables but ' v ' as its sole connective: (a) show that A cannot be tautologous; (b) show that A cannot be inconsistent.

4   Consider the three wffs:

   (a) $P \& (R \rightarrow R) \rightarrow -Q$

   (b) $P \& (Q \vee R)$

   (c) $P \rightarrow (Q \leftrightarrow R)$

      (i) Show that none of (a)–(c) implies any other.

      (ii) Show that one and only one of (a)–(c) is subcontrary to both the others.

5   (i) Show that, for any wffs A and B, if A and B are contrary then each implies the negation of the other and their negations are subcontrary.

   (ii) Show that, for any wffs A and B, if A and B are subcontrary then the negation of each implies the other and their negations are contrary.

   (iii) Show that, for any wffs A and B, A and B are equivalent (a) if and only if $-A$ and $-B$ are equivalent, and (b) if and only if A and $-B$ are contradictory.

6   (i) Draw up a table showing all possible truth-functions of *one* variable, and for each find an equivalent expression containing only ' $\rightarrow$ ' and ' $-$ '.

   (ii) How many distinct possible truth-functions of *three* variables are there? Of *n* variables?

4 THE CONSISTENCY OF THE PROPOSITIONAL CALCULUS

We apply now the truth-table approach to the propositional calculus in order to obtain an affirmative answer to the question: are our rules of derivation safe? We in fact show that every theorem provable from the rules is tautologous by a truth-table test, so that no contingent theorems and no inconsistent theorems exist, and in particular no contradiction A & −A, which would of course be inconsistent, can be derived. We are interested, however, not only in theorems but in sequents in general; we want to ensure that all derivable sequents are reliable patterns of argument. To achieve this broader result, we need first to extend the application of the truth-table procedure from wffs to sequent-expressions.

The extension is easy. Let

$$A_1, \ldots, A_n \vdash B$$

be any sequent-expression. Then a truth-table test on $A_1, \ldots,$ $A_n \vdash B$ is an evaluation of the wffs $A_1, \ldots, A_n$, B for every possible assignment of truth-values to the variables occurring in $A_1, \ldots,$ $A_n \vdash B$. We may display such a test by listing to the left of the sequent-expression all variables occurring in *any* of the wffs $A_1, \ldots,$ A_n, B, writing under them all possible assignments of truth-values in the standard manner, and then truth-table testing each wff in the sequent-expression separately for these assignments. Thus:

P	Q	$P \rightarrow Q,$	$-Q \vdash$	$-P$
T	T	T T T	F T	F T
T	F	T F F	T F	F T
F	T	F T T	F T	T F
F	F	F T F	T F	T F

Next, we define what it is for a sequent-expression to be tautologous. $A_1, \ldots, A_n \vdash B$ is *tautologous* if, for every assignment of truth-values to its variables for which all of $A_1, \ldots, A_n$ take the value T, B takes the value T also. The sequent-expression just tested is tautologous; there is only one assignment ($P =$ F, $Q =$ F) for which '$P \rightarrow Q$' and '$-Q$' both take the value T, and for this assignment the conclusion '$-P$' also takes the value T. Equivalently $A_1, \ldots, A_n \vdash B$ is tautologous if there is no assignment for which $A_1, \ldots, A_n$ are all true and B false. In the limiting case where the

75

sequent-expression has *no* assumptions $A_1, \ldots, A_n$, this definition simply requires that B take the value T for all assignments of truth-values to its variables, and so $\vdash$ B will be a tautologous sequent-expression just in case B is a tautologous wff.

A second, sometimes useful, way of explaining a tautologous sequent-expression is to associate with each sequent-expression $A_1, \ldots, A_n \vdash$ B a single wff which we call *the corresponding conditional*:

$$A_1 \rightarrow (A_2 \rightarrow (\ldots (A_n \rightarrow B) \ldots)).$$

Thus for the sequent-expression tested above the corresponding conditional is

$$(P \rightarrow Q) \rightarrow (- Q \rightarrow -P).$$

As a limiting case, if there are no assumptions $A_1, \ldots, A_n$, the corresponding conditional is to be simply B itself (which is slightly queer, since B may actually not be a conditional at all). Then we can show that a sequent-expression is tautologous (in the sense defined above) if and only if its corresponding conditional is tautologous (in the sense of the previous section).

For suppose that for a sequent-expression $A_1, \ldots, A_n \vdash$ B its corresponding conditional is not tautologous. Then some assignment of truth-values to the variables in $A_1, \ldots, A_n \vdash$ B yields F as the value of $A_1 \rightarrow (A_2 \rightarrow (\ldots (A_n \rightarrow B) \ldots))$. By the matrix for '$\rightarrow$', this is only possible if $A_1 = $ T, and $A_2 \rightarrow (\ldots (A_n \rightarrow B) \ldots) = $ F for that assignment. This in turn is only possible if $A_2 = $ T and $A_3 \rightarrow (\ldots (A_n \rightarrow B) \ldots) = $ F. Continuing, we see that for this assignment $A_1, \ldots, A_n$ must *all* take the value T and B the value F, whence $A_1, \ldots, A_n \vdash$ B is not tautologous either. Conversely, suppose $A_1, \ldots A_n \vdash$ B is not tautologous. Then for some assignment $A_1, \ldots, A_n$ all take the value T whilst B takes the value F. For this assignment, by the matrix for '$\rightarrow$', $A_n \rightarrow B = $ F, whence $A_{n-1} \rightarrow (A_n \rightarrow B) = $ F, and so on. Hence, the corresponding conditional, $A_1 \rightarrow (A_2 \rightarrow (\ldots (A_n \rightarrow B) \ldots))$, must take the value F for this assignment, and so is not tautologous either.

As an alternative, therefore, to the above test, we may test the corresponding conditional of a sequent-expression in the normal way to establish whether the sequent-expression is tautologous or not.

Let us, for brevity, call a sequent *derivable* if a proof can be found for it employing only the ten primitive rules of derivation. The main result of this section is that all derivable sequents are tautologous. We state the result in the following *metatheorem* (we have reserved the title ' theorem ' for certain results *in* the propositional calculus, but this is rather a result *about* the calculus):

Metatheorem I: All derivable sequents are tautologous.

Outline of proof. We should bear in mind that the number of derivable sequents is indefinitely large (for example, all possible substitution-instances of a derivable sequent are also derivable, as we know by (S2)), so that we cannot proceed by inspecting individual sequents in turn: we need a more general method of proof. The method we employ is closely related to the mathematical method known as *proof by induction*. If we wish to show that *all* numbers have a certain property, it suffices to show that 0 has the property, and that if a given number has the property then the next number in sequence has the property.[1] Given that 0 has the property, we can thus show that 1 has it; given that 1 has it, we can show that 2 has it; and so on, up to any given number. In our present case, remember that a derivable sequent has, by definition, a proof, and that this proof, which may be as long as we please but can only be finitely long, proceeds in stages. It has to begin with an application of the rule A, since there is no other way of initiating a proof. And any later step (unless it is also an application of A) is based in a definite way on earlier lines of the proof. If we can show, therefore, (i) that any application of A by itself yields a tautologous sequent; and (ii) that any application of the other nine rules based on lines corresponding to tautologous sequents yields also a tautologous sequent, then we shall effectively have shown that any sequent which is derivable at all is tautologous. Thus our proof falls into two stages: we show (α) that any application of A yields a tautologous sequent; we show (β) that, *if* at a given stage in a proof the earlier lines correspond to tautologous sequents, *then* an application of one of the other nine rules to some of these lines yields a resulting line which also corresponds to a tautologous sequent. Thus in (β) the work naturally falls into nine phases, corresponding to the nine rules, and in each phase we establish a *conditional proposition*.

---

[1] By ' number ' is here meant a natural number (cf. footnote 1, page 105), i.e. one of the numbers 0, 1, 2, 3, etc.

Proof of (α). Any sequent derivable by the rule of assumptions alone is tautologous. For such a sequent must be of the form A ⊢ A, which is obviously tautologous.

Proof of (β). (i) MPP. In an application of MPP, we pass from premisses A and A → B to conclusion B, on the pool of the assumptions on which A and A → B rest. Now suppose that the lines where A and A → B appear correspond to tautologous sequents. We wish to show that the new line where B appears will also correspond to a tautologous sequent. Let us suppose this is not so, and deduce an absurdity. If the new sequent is not tautologous, then clearly some assignment of truth-values to its variables gives B the value F but all its assumptions the value T. Since these assumptions include all those on which A and A → B rest, the same assignment will give A and A → B the value T, because by supposition the lines where they appear correspond to tautologous sequents. But if this assignment gives both A and A → B the value T, it must also give B the value T, by the matrix for ' → ': an absurdity, since we supposed B to take the value F for the assignment in question. Thus any application of MPP to tautologous sequents yields tautologous sequents.

(ii) DN. In an application of DN, we pass from premiss A to conclusion − −A on the same assumptions, or vice versa. Suppose, in the first case, that the line where A appears is tautologous. Then any assignment of truth-values to the variables in the sequent there proved which gives each assumption the value T gives A the same value T. By the matrix for ' − ', each such assignment will also give − −A the value T, so that the new line where − −A appears will also be tautologous. The second case (where − −A is the premiss and A the conclusion) is similar. Thus any application of DN to tautologous sequents yields tautologous sequents.

(iii) MTT. Considerations similar to those in (i) show that any application of MTT to tautologous sequents yields tautologous sequents. (Since MTT can be obtained as a derived rule, as was shown in Section 2 of this chapter, it need in any case not be considered here.)

(iv) CP. Effectively, in an application of CP, we pass from a sequent of the form $A_1, \ldots, A_n, A_{n+1} \vdash B$ to one of the form $A_1, \ldots, A_n \vdash A_{n+1} \to B$, where A_{n+1} is the discharged assumption. Suppose that $A_1, \ldots, A_n, A_{n+1} \vdash B$ is tautologous, but that $A_1, \ldots,$

$A_n \vdash A_{n+1} \rightarrow B$ is not. Then some assignment of truth-values to the variables in the sequent gives $A_1, \ldots, A_n$ all the value T, and $A_{n+1} \rightarrow B$ the value F. By the matrix for ' $\rightarrow$ ', this assignment gives A_{n+1} the value T and B the value F, for otherwise $A_{n+1} \rightarrow B$ would have the value T. But such an assignment would render $A_1, \ldots A_n$, $A_{n+1} \vdash B$ non-tautologous, contrary to supposition. Thus any application of CP to tautologous sequents yields tautologous sequents.

(v) &I. In an application of &I, we pass from premises A and B to conclusion A & B, on the pool of the assumptions on which A and B rest. Suppose that the lines where A and B appear correspond to tautologous sequents. Then any assignment of truth-values to the variables in the pool of assumptions which gives all these assumptions the value T gives A and B separately the value T, since their respective assumptions are included in the pool. By the matrix for ' & ', therefore, any such assignment gives A & B the value T also. Thus any application of &I to tautologous sequents yields tautologous sequents.

(vi) &E. It is left as an (easy) exercise for the reader to show that any application of &E to tautologous sequents yields tautologous sequents.

(vii) vI. Consideration of the ' v ' matrix shows that any application of vI to tautologous sequents yields tautologous sequents. Details are left to the reader.

(viii) vE. In an application of vE, we pass from premises A v B, together with a proof of C from A and a proof of C from B, to the conclusion C on the pool of the assumptions on which A v B rest and those used to derive C from A (apart from A) and those used to derive C from B (apart from B). Now suppose that the line where A v B appears corresponds to a tautologous sequent, and so do the lines where C appears as a conclusion from A and as a conclusion from B. We have to show that the line where C appears as a conclusion from the complex pool of assumptions is also a tautologous sequent. Suppose, therefore, that it is not. Then some assignment of truth-values to its variables gives all the assumptions the value T, but C the value F. These assumptions include all those on which A v B rests, whence the assignment must give A v B the value T. By the ' v ' matrix, therefore, either A or B separately has the value T for this assignment. Suppose that A has the value T. Then the line where C is derived from A cannot be tautologous, since this assignment gives all its assumptions including A the value T, yet C the value F. Suppose then that B has the value T. Similarly, the line where C is derived from B in this case cannot be tautologous. In either case, we have an absurdity. Hence any application of vE to tautologous sequents yields tautologous sequents.

(ix) RAA. Effectively, in an application of RAA, we pass from a sequent of the form $A_1, \ldots, A_n, A_{n+1}, \vdash B \& -B$ to one of the form $A_1, \ldots, A_n \vdash -A_{n+1}$. Suppose that the former sequent is tautologous but the latter not. Then some assignment of truth-values to the variables in the latter sequent gives $A_1, \ldots, A_n$ all the value T and $-A_{n+1}$ the value F, whence, by the matrix for ' — ', it must give A_{n+1} the value T. The assignment thus gives all the assumptions of the former sequent the value T, so that, on the supposition that this is tautologous, it must give B & —B the value T also, which is absurd by the matrices for ' & ' and ' — '. Thus any application of RAA to tautologous sequents yields tautologous sequents.

In sum, if any application of any of the nine rules is made on tautologous sequents the result is a tautologous sequent. Since by A we can only start with tautologous sequents, any sequent we can derive by our primitive rules is tautologous, and we have the metatheorem.

As an immediate consequence of Metatheorem I, for the special case where a derivable sequent has no assumptions and so its conclusion is a theorem, we have:

Corollary I: All theorems of the propositional calculus are tautologous.
We say that a logical system such as the propositional calculus is *consistent* if our rules for it do not enable us to derive as a theorem a contradiction. Since a contradiction A & —A is an inconsistency, and so not a tautology, we have from Corollary I:

Corollary II: The propositional calculus is consistent.
It should now be clear why, in discussing derived rules such as TI and SI, I stressed that a proof employing these rules could always be replaced by a (generally longer) proof of the same sequent employing only primitive rules. Otherwise we could not be sure,

without special proof, that these rules would not enable us to prove sequents which were non-tautologous.

Let us see why Metatheorem I should inspire confidence in the trustworthiness of our rules of derivation. A necessary condition of sound argument, we insisted at the outset, was that in it we may never pass from true assumptions to a false conclusion. Any argument which has as its pattern a derivable sequent of the propositional calculus will, we can now see, satisfy this condition at least. For any such sequent, by Metatheorem I, is tautologous, which is to say that in any case where its assumptions are true its conclusion is true also. There are no particular propositions by which we can replace the P, Q, R, etc., of a derivable sequent so as to obtain true assumptions but a false conclusion. In this way the truth-table approach has helped to answer a fundamental question about our rules of derivation.

To the extent that we now, it is hoped, feel a degree of confidence in the soundness of derivable sequents, we may use our rules to show the validity of actual arguments by establishing that their patterns are provable from the rules. The truth-table approach, however, enables us in addition to show the *invalidity* of argument-patterns. Suppose a truth-table test shows that a certain sequent is *not* tautologous. Then by Metatheorem I it is *not* derivable from our rules; nor would we wish it to be, for this test shows that for some assignment of truth-values to its variables we can verify all its assumptions yet falsify its conclusion. We have only to replace the values in this assignment by actual propositions with these values to discover an instance of the argument-pattern in question which is obviously invalid—true assumptions and a false conclusion. We can, therefore, replace the search for such instances, which may require imagination, by a truth-table test, which requires at worst only patience.

Often the full labour of a truth-table test can be replaced by a quick check. For example, consider

(1) $P \rightarrow (Q \rightarrow R), P \,\&\, -Q \vdash -R.$

Valid or invalid? Let us try to *invalidate* it, by assigning the value T to ' $P \rightarrow (Q \rightarrow R)$ ' and to ' $P \,\&\, -Q$ ', and the value F to ' $-R$ '. If $P \,\&\, -Q = $ T, then ' P ' and ' $-Q$ ' must both take value T, whence $P = $ T, $Q = $ F. If $-R = $ F, then $R = $ T. But for these values, we *do* get $P \rightarrow (Q \rightarrow R) = T$. In other words, the assignment $P = T$, $Q = F$, $R = T$ gives the value T to both assumptions and F to the conclusion, thus showing without the full eight-line test that the sequent is not tautologous. On the other hand, consider

(2) $P \rightarrow (Q \rightarrow R), P \& -R \vdash -Q.$

When we attempt to invalidate (2) by taking $P \& -R = T$, $-Q = F$, we obtain $P = T$, $R = F$, $Q = T$. But for this assignment $P \rightarrow (Q \rightarrow R) = F$. This shows that no assignment of truth-values can render both assumptions true and conclusion false; hence the sequent is tautologous, and we can proceed to seek a proof of it from our rules.

A final word about the paradoxes of material implication (compare Section 2): in view of the matrix for ' $\rightarrow$ ' given in Section 3, we rate a conditional $P \rightarrow Q$ as true if its consequent Q is true *whatever the truth-value of its antecedent P*: and we rate a conditional $P \rightarrow Q$ as true if its antecedent P is false *whatever the truth-value of its consequent Q*. The results 50 and 51 can be seen as simple reflections, in sequent form, of these facts, and can be accepted as safe patterns of reasoning given this understanding about the truth-value of ' $\rightarrow$ '. By Metatheorem I, they are tautologous, and so will not lead us from a true assumption to a false conclusion. In fact, ' $P \rightarrow Q$ ' differs from ' if P then Q ' (in the ordinary sense of these words) precisely to the extent that ' $P \rightarrow Q$ ' is a *truth-function* of P and Q—has its truth-value completely determined by those of P and Q—in a way that ' if P then Q ' is not. What they have in *common*, however, is the all-important feature that both are reckoned *false* in case P is true and Q false: it is a necessary condition for the truth of ' if P then Q ' that it is not both the case that P and not Q; but this condition is both necessary *and sufficient* for the truth of ' $P \rightarrow Q$ ', as its equivalence to ' $-(P \& -Q)$ ' reveals.

EXERCISES

1 As a further exercise in proof-discovery, and because we need the results in the next section, prove the following sequents:

(a) $P \& Q \vdash P \rightarrow Q$

(b) $-P \& Q \vdash P \rightarrow Q$

(c) $-P \& -Q \vdash P \rightarrow Q$

(d) $P \& -Q \vdash -(P \& Q)$

(e) $-P \& Q \vdash -(P \& Q)$

(f) $-P \& -Q \vdash -(P \& Q)$

(g) $P \& -Q \vdash P \lor Q$

(h) $-P \& Q \vdash P \lor Q$

(i) $P \& Q \vdash P \leftrightarrow Q$

(j) $P \& -Q \vdash -(P \leftrightarrow Q)$

(k) $-P \& Q \vdash -(P \leftrightarrow Q)$

(l) $-P \& -Q \vdash P \leftrightarrow Q$

2 Show the invalidity of the following patterns of argument by finding an assignment of truth-values to the variables such that the assumption(s) are all true and the conclusion false:

(a) $P \& Q \rightarrow R \vdash P \rightarrow R$

(b) $P \rightarrow Q \lor R \vdash P \rightarrow Q$

(c) $P \rightarrow Q, P \rightarrow R \vdash Q \rightarrow R$

(d) $P \rightarrow R, Q \rightarrow R \vdash P \rightarrow Q$

(e) $P \rightarrow (Q \rightarrow R), Q, -R \vdash P$

(f) $P \leftrightarrow -Q, Q \leftrightarrow -R, R \leftrightarrow -S \vdash P \leftrightarrow S$

5 THE COMPLETENESS OF THE PROPOSITIONAL CALCULUS

The attentive reader will no doubt have noticed that, though in the last section I showed that all derivable sequents were tautologous, it remains an open question whether the converse holds: whether, that is, all tautologous sequents are derivable from our rules. The burden of this section is to show that this is so. The importance of the result lies in the fact that it provides an answer to an outstanding question raised at the outset of the chapter: in what sense can we regard our rules as complete? In view of what we are about to prove, they are complete in the sense that they afford proofs for all tautologous sequents. Hence, if we added rules which enabled us to prove sequents not already derivable, these could only be non-tautologous sequents, which on intuitive grounds we should not wish to regard as sound patterns of argument: for there would be for them an assignment of truth-values making the assumptions all true and the conclusion false.

The proof that follows is rather involved. We prove as

Metatheorem II that all tautologous wffs (rather than sequents) are derivable as theorems. The proof of this proceeds by a *lemma*, or auxiliary result, which is required on the way. Metatheorem III, that all tautologous sequents are derivable, follows fairly easily from Metatheorem II. We finally draw certain corollaries from these metatheorems, and discuss the propositional calculus in the light of the results.

Metatheorem II: All tautologous wffs are derivable as theorems.

Outline of proof. We select a wff A which by hypothesis is tautologous under a truth-table test. A may, of course, be of great length and complexity; all we know of it is that for every assignment of truth-values to its constituent variables it takes the value T. Our task is to show how a proof of such a wff as a theorem, using only our ten rules of derivation, can in general be found. We do this essentially by imitating in the required proof the truth-table test itself. As our lemma we show that corresponding to each line of a truth-table test on *any* wff a derivable sequent can be written down; the assumptions of the sequent are the variables in the wff given, appearing either negated or non-negated according as the variable has the value F or T in the assignment in question; the conclusion of the sequent is the wff being tested, appearing either negated or non-negated according as it takes the value F or T for the assignment in question. We then show how, in the case of a *tautologous* wff A, we can use the derivable sequents yielded by the lemma to derive A as a theorem.

Proof.

Lemma. Let A be any wff, containing the propositional variables $V_1, \ldots, V_n$, and consider some assignment of truth-values to $V_1, \ldots, V_n$. For each such variable V_i, let W_i be either V_i or $-V_i$, according as V_i takes value T or F in the given assignment. Then we can derive either

$$W_1, \ldots, W_n \vdash A$$

or

$$W_1, \ldots, W_n \vdash -A,$$

according as A takes value T or F for this assignment.

Example. Suppose A is ' $-P \rightarrow -Q \vee R$ ', and consider the assignment $P = F$, $Q = T$, $R = F$. For this assignment, the wffs W_i are ' $-P$ ' (since $P = F$), ' Q ' (since $Q = T$), and ' $-R$ ' (since $R = F$). For this assignment, ' $-P \rightarrow -Q \vee R$ ' takes value F, as a test quickly reveals. Hence by the lemma the second alternative applies, and

$$-P, Q, -R \vdash -(-P \rightarrow -Q \vee R)$$

is a derivable sequent.

Proof of lemma. The technique of proof is related to that employed in the proof of Metatheorem I, and again resembles mathematical induction. We show: (α) that the lemma holds in case A is the shortest possible wff, namely a propositional variable; and (β) that *if* the lemma holds for wffs B and C, *then* it also holds for $-B$, $B \rightarrow C$, B & C, B $\vee$ C, and B $\leftrightarrow$ C. It follows from (α) and (β) that the lemma holds for *any* wff, since any wff is constructed, in virtue of the formation rules in Section 1, out of propositional variables by introducing in a systematic manner connectives such as ' $-$ ', ' $\rightarrow$ ', ' & ', ' $\vee$ ', and ' $\leftrightarrow$ '.

(α) Suppose that A is a propositional variable V. There are only two possible assignments to V, namely $V = T$ and $V = F$. Consider the assignment $V = T$; then A, being V, takes value T itself. We have to show, therefore, that

$$V \vdash V$$

is derivable. This is immediate by the rule A (cf. sequent 29). Similarly, if $V = F$, then A, being V, takes also value F, and we have to show that

$$-V \vdash -V$$

is derivable. This is immediate by rule A, as before.

(β) Suppose that the lemma holds for wffs B and C. We are to show that it holds for $-B$, $B \rightarrow C$, B & C, B $\vee$ C, and B $\leftrightarrow$ C. We consider the cases in turn.

(i) $-B$. Let $V_1, \ldots, V_n$ be the variables in B, and suppose, first, that for a given assignment B takes the value T. For this assignment $-B$ will take the value F. By hypothesis, the lemma applies to B, so that

$$W_1, \ldots, W_n \vdash B$$

is derivable, where $W_1, \ldots, W_n$ are related to $V_1, \ldots, V_n$ as described in the lemma. We have to prove that

$$W_1, \ldots, W_n \vdash --B$$

is derivable. But this is immediate from the proof of the given sequent, by one step of DN. Suppose, second, that for a given assignment B takes value F, whence $-B$ takes value T. By hypothesis, the lemma applies, so that

$$W_1, \ldots, W_n \vdash -B$$

is derivable. But this is exactly the sequent we need to prove to show that the lemma applies to $-B$.

(ii) $B \rightarrow C$. Let $U_1, \ldots, U_j$ be the variables in B and $V_1, \ldots, V_k$ the (not necessarily distinct) variables in C. We have to consider four cases, according as, for a given assignment to $U_1, \ldots, U_j$, $V_1, \ldots, V_k$, $B = T$ and $C = T$, $B = T$ and $C = F$, $B = F$ and $C = T$, or $B = F$ and $C = F$.

(*a*) Suppose that $B = T$ and $C = T$. Then, by the ' $\rightarrow$ ' matrix, $B \rightarrow C = T$. By hypothesis, we can derive

$$W_1, \ldots, W_j \vdash B$$

and

$$X_1, \ldots, X_k \vdash C,$$

where $W_1, \ldots, W_j$ are related to $U_1, \ldots, U_j$, and $X_1, \ldots, X_k$ to $V_1, \ldots, V_k$ as described in the lemma. We have to prove that

$$W, \ldots, W_j, X_1, \ldots, X_k \vdash B \rightarrow C$$

is derivable. Now from the proofs of the given sequents we can readily construct, by a step of &I together with a renumbering of lines where necessary, a proof of

$$W_1, \ldots, W_j, X_1, \ldots, X_k \vdash B \& C.$$

A step of SI, using 2.4.1(*a*) ($P \& Q \vdash P \rightarrow Q$), yields the desired sequent.

(*b*) Suppose that $B = T$ and $C = F$. Then $B \rightarrow C = F$. We have now as derivable

$$W_1, \ldots, W_j \vdash B$$

and

$$X_1, \ldots, X_k \vdash -C,$$

and we are to show that

$$W_1, \ldots, W_j, X_1, \ldots, X_k \vdash -(B \twoheadrightarrow C)$$

is derivable. A step of &I, as in (*a*), followed by SI using 2.2.5(*g*) ($P \And -Q \vdash -(P \twoheadrightarrow Q)$) suffices.

(*c*) Similar to (*a*) and (*b*), using SI with 2.4.1(*b*).

(*d*) Similar to (*a*) and (*b*), using SI with 2.4.1(*c*).

(iii) B & C. We consider four cases, as in (ii), and use the same notation.

(*a*) Suppose B = T and C = T. Then B & C = T. A step of &I suffices to obtain, from

$$W_1, \ldots, W_j \vdash B$$

and

$$X_1, \ldots, X_k \vdash C,$$
$$W_1, \ldots, W_j, X_1, \ldots, X_k \vdash B \And C.$$

(*b*) Suppose B = T and C = F. Then B & C = F. Employ &I and SI with 2.4.1(*d*) to obtain $-(B \And C)$ from $B \And - C$.

(*c*) Similar to (*b*), using SI with 2.4.1(*e*).

(*d*) Similar to (*b*), using SI with 2.4.1(*f*).

(iv) B v C. The four possible cases are covered in turn by using SI with 1.3.1(*e*), 2.4.1(*g*), 2.4.1.(*h*), and 1.5.1(*f*) respectively.

(v) B $\longleftrightarrow$ C. The four possible cases are covered in turn by using SI with 2.4.1(*i*)–(*l*).

(α) and (β) together show that we can, in our proofs, imitate the steps of the truth-table evaluation of a wff for a given assignment to its variables. Consider again the example that follows the statement of the lemma. In virtue of (α) we can derive all three sequents

$$-P \vdash -P$$
$$Q \vdash Q$$
$$-R \vdash -R.$$

For the given assignment, ' $-Q$ ' takes value F, whence by (β) (i) we can construct a proof of

$$Q \vdash - -Q.$$

For the given assignment, ' $-Q \text{ v } R$ ' takes value F, whence by (β) (iv) we can construct from the given proofs a proof of

$$Q, -R \vdash -(-Q \text{ v } R).$$

Finally, for the given assignment, ' $-P \rightarrow -Q \text{ v } R$ ' takes value F, whence by (β) (ii) we can construct from the given proofs a proof of

$$-P, Q, -R \vdash -(-P \rightarrow -Q \text{ v } R),$$

which is the sequent to be derived according to the lemma. Thus (α) and (β) together can be seen to yield the lemma in full generality.

In order to prove Metatheorem II from the lemma, let A now be a *tautologous* wff, containing the variables $V_1, V_2, \ldots, V_n$. Since A is tautologous, it takes value T for *all* assignments to its variables, whence, by the lemma, we can derive *all* sequents of the pattern

$$W_1, \ldots, W_n \vdash A,$$

where $W_1, \ldots, W_n$ are negated or non-negated versions of the variables $V_1, \ldots, V_n$. There will be 2^n such sequents, corresponding to the 2^n possible assignments.

Example. ' $P \text{ \& } Q \rightarrow P$ ' is a tautology, whence by the lemma we can derive all four sequents

(i) $P, Q \vdash P \text{ \& } Q \rightarrow P$

(ii) $P, -Q \vdash P \text{ \& } Q \rightarrow P$

(iii) $-P, Q \vdash P \text{ \& } Q \rightarrow P$

(iv) $-P, -Q \vdash P \text{ \& } Q \rightarrow P.$

To construct a proof of A as a theorem, we begin by n steps of TI, introducing successively $V_1 \text{ v } -V_1, V_2 \text{ v } -V_2, \ldots, V_n \text{ v } -V_n$ (Theorem 44). Then, preparatory to steps of vE, we assume V_1, $-V_1, V_2, -V_2, \ldots, V_n, -V_n$. By SI applied 2^n times, using the 2^n sequents obtained from the lemma, we obtain A as a conclusion from the various assumptions in those sequents. Finally, by a succession of steps[1] of vE, we eventually obtain A resting on *no* assumptions, i.e. as a theorem. This proves the metatheorem.

Example continued. In the case of '$P \,\&\, Q \rightarrow P$', the proof as described goes as follows:

	(1) $P \vee -P$	TI 44
	(2) $Q \vee -Q$	TI(S) 44
3	(3) P	A
4	(4) $-P$	A
5	(5) Q	A
6	(6) $-Q$	A
3,5	(7) $P \,\&\, Q \rightarrow P$	3,5 SI (i)
3,6	(8) $P \,\&\, Q \rightarrow P$	3,6 SI (ii)
4,5	(9) $P \,\&\, Q \rightarrow P$	4,5 SI (iii)
4,6	(10) $P \,\&\, Q \rightarrow P$	4,6 SI (iv)
5	(11) $P \,\&\, Q \rightarrow P$	1,3,7,4,9 vE
6	(12) $P \,\&\, Q \rightarrow P$	1,3,8,4,10 vE
	(13) $P \,\&\, Q \rightarrow P$	2,5,11,6,12 vE

It should be clear from this example[2] why in general the steps of vE gradually reduce the number of assumptions on which A rests until, when the last substitution-instance of the excluded middle law is used, there are no assumptions left at all.

Metatheorem III: All tautologous sequents are derivable.

Proof. Let $A_1, \ldots, A_n \vdash B$ be a tautologous sequent. Then we know that its corresponding conditional

$$A_1 \rightarrow (\ldots (A_n \rightarrow B) \ldots)$$

is also tautologous (see Section 4). Hence by Metatheorem II this conditional can be derived from our rules as a theorem. We can derive the given sequent, therefore, as follows. First, assume $A_1, \ldots, A_n$. By TI, introduce the corresponding conditional as a

---

[1] How many such steps in general? The keen student should confirm that $2^{n-1} + 2^{n-2} + \ldots + 1$ steps will suffice.

[2] To avoid a possible misunderstanding: of course '$P \,\&\, Q \rightarrow P$' is easily proved without all this fuss; it is simply taken here as an *illustration* of a tautology in two variables 'P' and 'Q'—any other, however long and cumbersome, would have done as well.

new line. By n steps of MPP, we obtain B as conclusion from $A_1, \ldots, A_n$, thus deriving the desired sequent.

Corollary I: Any sequent is derivable if and only if it is tautologous. This Corollary simply puts together Metatheorems I and III.

We say that a logical system such as the propositional calculus is *complete* if all expressions of a specified kind are derivable in it. If, in particular, we specify *tautologous* sequent-expressions, then from Metatheorem III we have at once

Corollary II: The propositional calculus is complete.

Comparing the last section with the present one, we might observe that a *consistency* result, such as was obtained in Section 4, is typically to the effect that *only* expressions of a certain kind are derivable (in the case of the propositional calculus, only tautologous sequent-expressions), whilst a *completeness* result, such as has just been obtained, is typically to the effect that *all* expressions of a certain kind are derivable. Put together, as in Corollary I above, we see that our rules of derivation do not go too far, but that they do go far enough: our rules enable us to derive *exactly* the tautologous sequents.

So the two very different approaches to the propositional calculus, the derivational approach and the truth-table approach, coincide after all. By truth-tables, we separate tautologous sequents from the rest; by derivation, we separate derivable sequents; either way, however, we finish with the same totality of sequents. This may raise doubts as to the utility of the derivational approach. For, as we have said, a truth-table test is entirely mechanical: why concern ourselves with the search for proofs when the same job can be done by mechanical means?

There are several replies we can make. First, even at the propositional calculus level, in practice the truth-table method becomes cumbersome for four or more variables; it is often easier to search for a proof, when we suspect a sequent to be tautologous. Second, the lines of a proof follow in many cases the methods of reasoning which we ordinarily and unthinkingly apply (for this reason, our derivational approach is sometimes called *natural deduction*),whilst a truth-table test has a somewhat artificial character. Third, if we have any grounds for wishing to *reject* a tautology as not a logical principle (for example, the law of excluded middle has often been questioned by philosophers), then an explicit proof from our rules reveals what has to be abandoned and what may be retained in view of such a rejection, which a truth-table test does not. Our approach by rules shows something of the *interdependence* of results, whilst a truth-table test is applied separately to each sequent-expression.

At more complex levels of logic, however, such as that of the predicate calculus to which we turn in the next two chapters, the truth-table approach breaks down; indeed there is known to be no mechanical means for sifting expressions at this level into the valid and the invalid. Hence we are *required* there to use techniques akin to derivation for revealing valid sequents, and we shall in fact take over the rules of derivation for the propositional calculus, expanding them to meet new needs. The propositional calculus is thus un-typical: because of its relative simplicity, it can be handled mech-anically—indeed, in view of Metatheorems II and III we can even generate proofs mechanically for tautologous sequents. For richer logical systems this aid is unavailable, and proof-discovery becomes, as in mathematics itself, an imaginative process.

CHAPTER 3

The Predicate Calculus 1

1 LOGICAL FORM: 'ALL' AND 'SOME'

Our ambition (compare Section 1 of Chapter 1) is to state exactly the conditions of valid argumentation; to inspire the reader with the correct degree of humility, it is now time to see how very little we have so far achieved.

Consider the two arguments

> (1) If he is at home, his hat will be in the hall; his hat is not in the hall; therefore he is not at home.

> (2) If Napoleon was Chinese, he was Asiatic; he was not Asiatic; so he was not Chinese.

With the aid of the propositional calculus, we see at once that both arguments are sound since they have as a common pattern

> (3) $P \rightarrow Q, -Q \vdash -P,$

which we can show to be a derivable sequent. Many of the sequents so far derived exhibit in this way the pattern, or logical form, of quite familiar, everyday arguments. But there are many, equally familiar, arguments which are undoubtedly sound but whose soundness is not revealed at all by our present methods. I remind the reader of two arguments given at the outset of the book:

> (4) Tweety is a robin; no robins are migrants; therefore Tweety is not a migrant.

> (5) Oxygen is an element; no elements are molecular; therefore oxygen is not molecular.

We agreed that, like (1) and (2), (4) and (5) have something in common which we called their logical form; but by the propositional calculus notation the only sequent we can write down for them is

> (6) $P, Q \vdash R,$

a patently invalid sequent.

It is worth comparing (1) and (2) with (4) and (5). In (1) and (2) we recognize the conclusion of the arguments as the negation of the proposition appearing as antecedent of the first premiss, and we recognize the second premiss as the negation of the consequent of the first; it is this recognition that yields the pattern (3). But in (4) and (5) there is no such re-identification of propositions, the two premisses and the conclusion are all distinct, so that all we can write down is (6). Inasmuch as (4) and (5) are sound, it is in virtue of the *internal structure* of the propositions that they are so, whilst to show the soundness of (1) and (2) it suffices to consider the propositions as units without looking inside them. The propositional calculus reveals validity when this depends on propositional structure *alone*; our units, at this level, are propositions themselves. Clearly what we need now are tools which enable us to break into propositions, to reveal their inner structure, if we are to pursue further the search for validity conditions. These tools are provided by the *predicate calculus*, the subject of this and the following chapter.

When we tackled (4) and (5) earlier, we reached the following partial analysis of their common logical form:

> (7) m has property F; nothing with F has property G; there-fore m does not have G.

We clearly shall need in our new formal language special symbols to replace proper names (' Tweety ', the proper name of a bird; ' oxygen ', the proper name of a chemical element). Let us employ ' m ', ' n ', . . ., for this purpose, as in (7). Also we shall need symbols to replace property-expressions or predicates (' is a robin ', ' is a migrant ', ' is an element ', ' is molecular '). Let us employ capitals ' F ', ' G ', ' H ', . . ., again as in (7). ' m ', ' n ', . . ., we call *proper names*, ' F ', ' G ', . . ., we call *predicate-letters*.

To say that m has property F, we agree to juxtapose the symbols ' F ' and ' m ' in that order. We write

$$Fm.$$

' Fm ' thus becomes our pattern for the many simple sentences which have proper names as subjects: ' Socrates is mortal ', ' Napoleon was Chinese ', ' courage is desirable ', and so on.

To say that m does not have G is now easy: for to say that m has G, we write ' Gm ', hence for ' m does not have G ' we write

$$-Gm,$$

borrowing, as we shall freely allow ourselves to do, symbols from the propositional calculus already developed.

This caters for the first premiss and for the conclusion of (7), but we are left without symbols for the expression 'nothing' in the second premiss. We could simply introduce a symbol (say ' E ') for this, and write ' E(*FG*) ' for ' nothing with *F* has *G* '; more or less this tactic was adopted by traditional logic, stemming from Aristotle.[1] In contemporary logic, a more flexible and subtle device is used, which is, however, harder to learn. We approach the problem obliquely by examining first ' everything ' rather than ' nothing '.

From ' no robins are migrants ' we obtained ' nothing with the property of being a robin has the property of being a migrant '. Similarly, from ' all robins are migrants ' or ' every robin is a migrant ' we obtain ' everything with the property of being a robin has the property of being a migrant '. Using ' *F* ' and ' *G* ' as before, we have

(8) Everything with *F* has *G*.

The step we must now take is to see that (8) amounts to a kind of *conditional*, namely:

(9) Everything, *if* it has *F*, has *G*,

or, perhaps better,

(10) Take anything you like: then if it has *F*, it has *G*.

To say that everything which has *F* has *G* is to say that, *for any object whatsoever*, if it has *F* it has *G*. To say that all robins are migrants is to say that, pick what object you will, if it is a robin then it is a migrant.[2]

We can render (9) and (10) symbolically by adopting from algebra the convenient device of *variables*, ' *x* ', ' *y* ', ' *z* ', ... In place of (9), we write first

---

[1] For a fuller discussion, see Chapter 4, Section 4.

[2] Here is a ' proof '. Suppose, first, that all robins are migrants, and arbitrarily choose some object, say a haystack. Then it is true of the haystack that *if* it is a robin it is a migrant. Conversely, suppose that not all robins are migrants. Then some robin is not a migrant. Pick one: then for this object at least it is not true that if it is a robin it is a migrant. In this way we see in general that (8) is true and false in exactly the same circumstances as (9) and (10).

(11) For any x, if x has F then x has G.

' x has F ' we naturally abbreviate to ' Fx ', ' x has G ' similarly to ' Gx ', and for ' if . . . then . . .' we use ' $\rightarrow$ ' as before. This gives

(12) For any x: $Fx \rightarrow Gx$.

We finally agree to write ' for any x ' by enclosing ' x ' in brackets, and obtain, as a fully symbolic version of (8),

(13) $(x)(Fx \rightarrow Gx)$.

The full advantages of adopting the device of variables will become clearer as we proceed. For the moment it will suffice to think of them as operating somewhat like the pronoun ' it '. When we compare (11) with (10), we observe in fact that ' x ' in (11) has twice replaced ' it ' in (10). The abbreviation ' (x) ' for ' for any x ' which appears in (13) is called a *universal quantifier*.

So ' Everything with F has G ' becomes ' $(x)(Fx \rightarrow Gx)$ '. Now, what of ' nothing with F has G '? Reflection shows that to say that *nothing* with F has G is just to say that *everything* with F *lacks* G, or, in explicitly conditional form,

(14) Take anything you like: then if it has F it does *not* have G.

This becomes in turn

(15) For any x: $Fx \rightarrow -Gx$

and

(16) $(x)(Fx \rightarrow -Gx)$,

which last is in full symbolic dress.

As a more complex example, suppose we wish to analyse

(17) No men are both doctors and fishmongers.

To obtain the version in terms of properties, we first write

(18) Nothing with the property of being a man has the property of being both a doctor and a fishmonger.

Let F be the property of being a man, G the property of being a doctor, and H the property of being a fishmonger. Then we obtain

(19) Nothing with *F* has both *G* and *H*,

and so

(20) Take anything you like: then if it has *F* it does not have both *G* and *H*,

whence

(21) For any x: $Fx \rightarrow -(Gx \,\&\, Hx)$,

whence

(22) $(x)(Fx \rightarrow -(Gx \,\&\, Hx))$.

The device of the universal quantifier enables us to render into logical notation many sentences, as they occur in arguments, which contain such words as ' all ', ' every ', ' any ', ' everything ' and also ' no ', ' none ', ' nothing '. Another group of idioms, however, plays a large role in the reasoning situation, a group centering on ' some '. Consider, for example, the sound argument

(23) All Germans are hilarious; some felons are German; therefore some felons are hilarious.

If we put ' *F* ' for being a felon, ' *G* ' for being German, ' *H* ' for being hilarious, then the first premiss evidently is expressed by ' $(x)(Gx \rightarrow Hx)$ '. It is unclear, however, how we should handle the second premiss and the conclusion. We may reflect that to say that some felons are German is to say that it is not the case that no felons are German. If ' no felons are German ' becomes ' $(x)(Fx \rightarrow -Gx)$ ', then ' it is not the case that no felons are German ' becomes ' $-(x)(Fx \rightarrow -Gx)$ '. Hence we *can* handle ' some ' with the aid of the universal quantifier again.

Though this analysis is perfectly correct, it proves more convenient for managing arguments to employ a special symbol in dealing with ' some ': the *existential quantifier*.

Just as

$$' (x)(...) '$$

is to mean ' take any x: then ...', so we write

$$' (\exists x)(...) '$$

to mean ' there is an x such that . . .', or ' an object x can be found which . . .'. ' $(\exists x)(. . .)$ ' affirms that at least one thing is such that . . . . Now to say that some felons are German is to say that at least one object can be found such that it has *both* property F *and* property G.[1] In general, to say that

(24) something with F has G

is to say that something has both F and G, and so becomes

(25) there is an x such that x has F and x has G,

or, in full symbolic dress,

(26) $(\exists x)(Fx \ \& \ Gx)$.

Another form of words common enough to deserve special mention is ' something with F has not G ' or ' there is something with F but not G ' (' some Frenchmen are not generous ', ' there are Frenchmen who are not generous '). This becomes, evidently, ' there is an x such that x has F and x has not G ', or

(27) $(\exists x)(Fx \ \& \ -Gx)$.

It is a common mistake to render ' some Frenchmen are generous ' by ' $(\exists x)(Fx \rightarrow Gx)$ ' rather than the correct ' $(\exists x)(Fx \ \& \ Gx)$ ', by assimilation to the case of ' all Frenchmen are generous ', which is properly rendered as a kind of conditional. But ' $(\exists x)(Fx \rightarrow Gx)$ ' affirms that there is something which, *if* it is French, then it is generous, and this will be true even if there are *no* Frenchmen, which ' some Frenchmen are generous ' certainly is not.

The task of translation into the quantifier-notation might be summarized thus: first, render into a sentence about properties, and employ predicate-letters for these properties; second, introduce variables; third, introduce propositional calculus connectives and quantifiers. The four commonest forms, after the first step is complete, together with their final translations, are exhibited in the following table:

[1] Always, in logic, we take ' some ' to mean ' at least one '. Thus ' some felons are German ' will count as true if there is only one German felon, and also true if all felons are German. See Chapter 4, Section 3, for the development of a device for expressing ' at least two '.

Everything with F has G	Nothing with F has G
$(x)(Fx \rightarrow Gx)$	$(x)(Fx \rightarrow -Gx)$
Something with F has G	Something with F has not G
$(\exists x)(Fx \,\&\, Gx)$	$(\exists x)(Fx \,\&\, -Gx)$

Traditional logic recognized these four forms of proposition, but no more. Its greatest limitation (apart from perhaps the virtual non-recognition of the propositional calculus) is just this lack of expressive power for other varieties, and the main merit of the quantifier-notation combined with variables and predicate-letters is that it enormously increases our powers of sentence-analysis. In the first place, we can handle *relations* as well as properties. Consider ' Prince Philip is a parent of Prince Charles '. Unlike ' Prince Philip is generous ', say, which we render ' Gm ' (using ' m ' for ' Prince Philip '), the new sentence contains two names, and a relation is affirmed between two objects rather than a property affirmed of one. Let us use ' n ' for ' Prince Charles ': then we can write ' Pmn ' as shorthand for ' m is parent of n '. ' Pmn ' will be true, but ' Pnm ' (' Prince Charles is a parent of Prince Philip ') false, so that when two proper names follow a predicate-letter order becomes material.

A predicate-letter followed by one name expresses a property; a predicate-letter followed by two names expresses a relation. Examples of relational sentences in ordinary speech which we can handle this way are: ' m loves n ', ' m is bigger than n ', ' m is to the south of n ', ' m is a second cousin of n '. Transitive verbs (at least when followed by objects), adjectives in the comparative form, expressions of relative position, and terms for family relationships, among many others, frequently express relations or occur in relational expressions.

In the second place, we can combine this extended use of predicate-letters with quantifiers. Let us suppose, for the moment, that our variables ' x ', ' y ', ' z ' range over *people* (as in algebra they are understood to range over numbers), so that for ' everything ' we can read ' everyone ' and for ' something ' ' someone '. Then, giving ' Pmn ' the interpretation ' m is a parent of n ', we can express ' Prince Charles has a parent ' by

(28) $(\exists x)Pxn$

where n is Prince Charles; this merely says that someone is a parent of Prince Charles. To say that Prince Philip has a child, we write

(29) $(\exists x)Pmx,$

where m is Prince Philip, which says that there is someone of whom Prince Philip is a parent. Again, order is essential here, because with the same interpretation, $(\exists x)Pnx$ (' there is someone of whom Prince Charles is a parent ') is, at the date of writing, false, whilst (28) is true. Similarly, $(\exists x)Pxm$, though true like (29), concerns the ancestry of Prince Philip rather than his progeny.

In the third place, we can combine universal and existential quantifiers in the same sentence. ' $(\exists x)Pxn$ ' says, of Prince Charles, that he has a parent. Suppose we wish to say that *everyone* has a parent. We naturally write

(30) $(y)(\exists x)Pxy$

(' take any y: then there is an x such that x is parent of y '). This must of course be carefully distinguished from

(31) $(y)(\exists x)Pyx,$

which affirms rather that everyone is a parent of someone, a generalization to which there are obvious exceptions. It is also essential to observe that the order of *quantifiers* is here as important to sense as the order of ' x ' and ' y '. Thus

(32) $(\exists y)(x)Pyx$

affirms with evident falsity that someone is a parent of everyone, and

(33) $(\exists y)(x)Pxy$

affirms that someone has everyone as a parent, which is even more obviously false.

In these last four examples, the use of *distinct* variables ' x ' and ' y ' is essential for the required sense. Thus to say ' $(\exists x)Pxx$ ' would be to say that someone was his own parent, and to say ' $(x)Pxx$ ' would be to say that everyone was his own parent. To express (30) without using variables, we should need ' for any person there is a person such that he (the latter) is a parent of him (the former) '; distinct variables enable us in a simple way to handle 'the former' and 'the latter', which again suggests a similarity between variables and pronouns.

There is no reason for stopping at *two* names or variables after a predicate-letter. To render, for example, 'Oxford is between London and Stratford', using '*m*' for 'Oxford', '*n*' for 'London', '*o*' for 'Stratford', and '*B*' for the relation of betweenness, we can write *Bmno*. In fact, in this book we shall rarely be concerned with relations involving more than two objects. But theoretically our predicate-letters may be followed by any (finite) number of names or variables.

I end this section with further illustrations of quantifier-translation in progressively more complex cases.

Consider

(34) Every boy loves a certain girl.[1]

We detect here an ambiguity: this may mean (i) that there is some one (very fortunate) girl who is loved by every boy; or (ii) that for every boy there can be found some (with luck, different) girl whom he loves. We obtain a different rendering for each version (a further merit of the quantifier-notation is that it makes explicit just this kind of ambiguity). Use '*B*' for the property of being a boy, '*G*' for the property of being a girl, and '*L*' for the loving relation. Then (i) becomes 'there is an *x* such that *Gx* and *x* is loved by every boy'. To render '*x* is loved by every boy', we think of it as 'every boy loves *x*'—'for all *y*, if *y* is a boy then *y* loves *x*'. Thus, putting together the pieces, we have

(35) $(\exists x)(Gx \mathbin{\&} (y)(By \to Lyx))$.

On the other hand (ii) becomes 'for all *x*, if *Bx* then *x* loves some girl', and to say '*x* loves some girl' we say 'there is a *y* such that *Gy* and *x* loves *y*'. Hence we have for (ii)

(36) $(x)(Bx \to (\exists y)(Gy \mathbin{\&} Lxy))$.

The overall structure of (i) is 'something with *G* has *H*'; but *H* is here the complex property of being loved by every boy. The overall structure of (ii), however, is 'everything with *B* has *H*', where *H* is now the complex property of loving some girl. (35) and (36) should be carefully compared and contrasted.

[1] This attractive example is due to Mr Peter Geach.

Consider next

(37) All the nice girls love a sailor.

Another ambiguity is detectable here: does this mean (i) that, for all x, if x is a nice girl then x loves *some* sailor, or (ii) that, for all x, if x is a nice girl then x loves *any* sailor? If (i), then, using ' N ' for being nice, ' G ' for being a girl, and ' S ' for being a sailor, we have

(38) $(x)(Nx \mathbin{\&} Gx \rightarrow (\exists y)(Sy \mathbin{\&} Lxy))$.

If (ii), on the other hand, we have

(39) $(x)(Nx \mathbin{\&} Gx \rightarrow (y)(Sy \rightarrow Lxy))$.

(Actually, I suspect that (37) means (ii).)

The occurrence of a conjunction as the antecedent of a conditional, as in (38) and (39), is typical of the analysis of ' all '-sentences with relative clauses attached to their grammatical subjects. Thus ' everything with F which has G has H ' becomes

(40) $(x)(Fx \mathbin{\&} Gx \rightarrow Hx)$.

' Except '-phrases often require a similar treatment. Thus ' all dogs except chihuahuas like the cold ' will become

(41) $(x)(Dx \mathbin{\&} -Cx \rightarrow Lx)$,

where ' D ' is for being a dog, ' C ' is for being a chihuahua, and ' L ' is for liking the cold. For ' dogs except chihuahuas ' means ' dogs *who are not* chihuahuas '.

The word ' any ' should always dictate caution. At the beginning of a sentence ' any ' plays usually the part of ' all ', and can be handled accordingly. But ' John does not like any girls' means, of course, ' John likes *no* girls ', and becomes

(42) $(x)(Gx \rightarrow -Lmx)$,

where ' G ' is for being a girl, ' L ' for the liking relation, and ' m ' is for John.

Careful rethinking is usually required when ' only ' is present. To say that only men are whisky-drinkers is to say that, for any x, only if x is a man is x a whisky-drinker, which is to say that, for any x, if x is a whisky-drinker then x is a man (' only if P then Q ' is equivalent to ' if Q then P '), which of course is to say that all whisky-drinkers are men. In general, ' Only things with F have G ' means ' Everything with G has F '. Hence, ' only men who eat nut steaks are vegetarians ' becomes

(43) $(x)(Vx \rightarrow Mx \mathrel{\&} Ex)$,

where ' V ' is for being vegetarian, ' M ' for being a man, and ' E ' for eating nut steaks: for it means the same as ' all vegetarians are men who eat nut steaks '. On the other hand ' *the only* men who eat nut steaks are vegetarians ' means ' *all* men who eat nut steaks are vegetarians ', and becomes

(44) $(x)(Mx \mathrel{\&} Ex \rightarrow Vx)$.

As much as this hinges on the simple presence of ' the '.

Flexibility of mind is generally required for translating from ordinary speech sentences into sentences of the predicate calculus. No firm rules can be given, and practice is needed before full familiarity with quantifiers is reached. The activity involved *is* one of translation; but the formal language into which translation is being made has a rather different syntax from that of a natural language, and has only a narrow terminology—logical connectives, variables, proper names, predicate-letters, and two quantifiers. The chief merit of the language is that, despite its notational limitations, it has a very wide expressive power: just how wide will become clearer as we proceed.

A final piece of terminology which will simplify some of the ensuing discussion: let us agree to call a proposition affirming that for any object such-and-such is the case a *universal proposition*; thus a universal proposition will be expressed in the predicate calculus by a sentence with an initial universal quantifier; and let us agree to call a proposition affirming that there is an object for which such-and-such is the case an *existential proposition*; thus an existential proposition will be expressed in our symbolism by a sentence with an initial existential quantifier.

EXERCISE

1 Exhibit the logical form of the following sentences by translating them into the notation of the predicate calculus (using for predicate-letters and proper names the letters suggested):

(*a*) Susan is featherbrained   ('*F*', '*m*')

(*b*) Janet is featherbrained   ('*F*', '*n*')

(*c*) Some women are featherbrained   ('*W*', '*F*')

(*d*) All women are featherbrained   ('*W*', '*F*')

(*e*) Only women are featherbrained   ('*W*', '*F*')

(*f*) No man is featherbrained   ('*M*', '*F*')

(*g*) Some men are not featherbrained   ('*M*', '*F*')

(*h*) John is not featherbrained   ('*F*', '*m*')

(*i*) Brutus killed Caesar   ('*K*', '*m*', '*n*')

(*j*) Someone killed Caesar   ('*K*', '*n*')

(*k*) Brutus killed someone   ('*K*', '*m*')

(*l*) Someone killed someone   ('*K*')

(*m*) Someone killed himself   ('*K*')

(*n*) No one killed himself   ('*K*')

(*o*) Someone killed everyone   ('*K*')

(*p*) Someone was killed by everyone   ('*K*')

(*q*) There is a town between London and Stratford   ('*T*', '*B*', '*m*', '*n*')

(*r*) Every woman owns a dog   ('*W*', '*D*', '*O*')

(*s*) Some dogs like every woman   ('*D*', '*W*', '*L*')

(*t*) Every featherbrained woman owns a dog   ('*F*', '*W*', '*D*'. '*O*')

(*u*) Every woman owns a featherbrained dog   ('*W*', '*F*', '*D*', '*O*')

(*v*) Fido likes a featherbrained woman   ('*m*', '*F*', '*W*', '*L*')

(*w*) Some dogs like a featherbrained woman   ('*D*', '*F*', '*W*', '*L*')

(*x*) Some dogs like only featherbrained women   ('*D*', '*F*', '*W*', '*L*')

(*y*) Some dogs do not like any featherbrained women   ('*D*', '*F*', '*W*', '*L*')

(*z*) Some dogs like only women who are not featherbrained   ('*D*', '*F*', '*W*', '*L*')

Warning: Some of the later sentences are ambiguous and need alternative renderings.

2 THE UNIVERSAL QUANTIFIER

The last section affords a preliminary sketch of a new formal language; we can now turn to the matter of testing arguments expressed in it. Since part of our language is just the propositional calculus itself, we take over into the predicate calculus the propositional connectives and propositional variables if we need them —all our earlier rules continue to be of service under the understanding that they are extended to the new symbolism. But we need additional rules for the handling of quantifiers in argument: four such, in fact—an introduction and an elimination rule for the universal and for the existential quantifier. We consider the universal quantifier first.

The elimination rule for the universal quantifier is concerned with the use of a universal proposition as a *premiss* to establish some conclusion, whilst the introduction rule is concerned with what is required by way of premiss for a universal proposition as *conclusion*. It is helpful to bear in mind the corresponding rules for ' & ', for there is a close similarity between ' & ' and the universal quantifier, as the following remarks suggest.

In particular arguments involving quantifiers, we usually have a particular group of objects, called our *universe of discourse*, in mind. For example, in algebra the variables ' x ', ' y ', ' z ', . . . are understood to range over numbers, so that our universe of discourse here is the set of all numbers; and, in discussing (28)–(33) in the last section, we explicitly restricted our universe of discourse to the set of all people. Our universe of discourse, in fact, is generally the understood range of our variables ' x ', ' y ', ' z ', ...

By way of illustration, let us suppose that our universe of discourse contains exactly 3 objects (what they are will not matter) whose proper names are ' m ', ' n ', and ' o '. Then to affirm that everything has property F will, for this universe, be to affirm that m has F and n has F and o has F. Thus

(1) $(x)Fx$

is intuitively equivalent, in this universe, to the complex conjunction with 3 conjuncts

(2) $Fm \,\&\, Fn \,\&\, Fo.$

Now by an obvious extension of &E, we could naturally derive as conclusion from (2) any one of the conjuncts separately, *Fm, Fn, Fo*. Analogously, our rule of universal quantifier elimination (UE) will allow us to infer that *any particular object* has *F* from the premiss that *all* things have *F*. The rule can be seen as a natural extension of &E, when we realize that affirming a proposition such as $(x)Fx$ is generally a condensed way of affirming a complex conjunction.

In fact, if all objects in a given universe had names which we knew and there were only finitely many of them, then we could always replace a universal proposition about that universe by such a complex conjunction. It is because these two requirements are not always met that we need universal quantifiers. For example, we may wish to say that all natural numbers [1] have a certain property *F*; this amounts to saying that 0 has *F*, and 1 has *F*, and 2 has *F*, and so on; but, there being infinitely many numbers, we are barred from actually completing the desired conjunction, and we fall back on the quantifier to do the job. Because our universe of discourse may be infinite in size, we cannot say that a universal proposition is *equivalent* to a complex conjunction, but it is true that the analogy with ' & ' is intuitively very helpful.

Hence the justification for UE is that, if everything has a certain property, any particular thing must have it, and UE will enable us to pass from $(x)Fx$ to conclusions such as *Fm* and *Fn*, and from $(x)(Fx \rightarrow Gx)$ to $Fm \rightarrow Gm$ and $Fn \rightarrow Gn$ (if everything is such that it has *G* if it has *F*, then in particular *m* has *G* if *m* has *F*, *n* has *G* if *n* has *F*). The rule is exemplified in the proof of the following elementary sequent:

100 $Fm, (x)(Fx \rightarrow Gx) \vdash Gm$

1	(1) Fm	A
2	(2) $(x)(Fx \rightarrow Gx)$	A
2	(3) $Fm \rightarrow Gm$	2 UE
1,2	(4) Gm	1,3 MPP

[1] By the *natural numbers* are meant the numbers 0, 1, 2, 3, etc. They are sometimes called also the *non-negative integers*, the *positive integers* being the numbers 1, 2, 3, etc.

100 exhibits the form of such obviously sound arguments as the logically famous

> (3) Socrates is a man; all men are mortal; therefore Socrates is mortal

(letting ' m ' be Socrates, ' F ' be being a man, and ' G ' be being mortal). We are now also in a position to validate the Tweety and oxygen arguments ((4) and (5) of the last section). Their common form (compare (7) of the last section) is proved as the following sequent:

101 $Fm, (x)(Fx \rightarrow -Gx) \vdash -Gm$

1	(1) Fm	A
2	(2) $(x)(Fx \rightarrow -Gx)$	A
2	(3) $Fm \rightarrow -Gm$	2 UE
1,2	(4) $-Gm$	1,3 MPP

The application of UE at line (3) is exactly like its application at the same line in the proof of 100: if everything with F lacks G, then in particular if m has F m lacks G.

The rule of universal quantifier introduction (UI) is designed for establishing as conclusions universal propositions. By the analogy with ' & ', to establish, say for our earlier universe of 3 objects, that everything has F, we should establish first that m has F, that n has F, and that o has F. Then, by an obvious extension of &I, we are sure that everything has F. This technique will be of no avail, however, if our universe is infinitely large or if we do not have names for all objects in the universe. We evidently require a new device.

Think of what Euclid does when he wishes to prove that all triangles have a certain property; he begins ' let ABC be a triangle ', and proves that ABC has the property in question; then he concludes that *all* triangles have the property.[1] What here is ' ABC '? Certainly not the *proper name* of any triangle, for in that case the conclusion would not follow. For example, given that Khrushchev is bald, it does not follow that everyone is bald. It is natural to view ' ABC ' as the name of an *arbitrarily selected triangle*, a particular triangle certainly but any one you care to pick. For if we can show

---

[1] See, for example, Euclid, *The Elements*, I, Propositions 16-21.

that an arbitrarily selected triangle has F, then we can soundly draw the conclusion that all triangles have F.

We introduce, therefore, the letters ' a ', ' b ', ' c ', . . . to be names (*not* proper names) of arbitrarily selected objects in the universe of discourse, and call them for short *arbitrary names*. Then, with important reservations to be made later, if we can show that Fa (an arbitrarily selected object has F) then we can conclude that $(x)Fx$.

In effect, a proof of Fa is tantamount to a proof of all the required conjuncts in the desired 'conjunction' $(x)Fx$. In the 3-object universe above, to prove Fa is to prove Fm, Fn, and Fo. For we can take m as the arbitrarily selected a, and n, and o. In the case where the universe is infinitely large, proving Fa is tantamount to proving infinitely many conjuncts, for we can select as a any object in the universe.

Hence the justification for UI is that, with certain reservations, if an arbitrarily selected object can be shown to have a property, everything must have it, and UI will enable us to pass from premises such as Fa or Fb to conclusion $(x)Fx$, and from $Fa \rightarrow Ga$ or $Fb \rightarrow Gb$ to $(x)(Fx \rightarrow Gx)$ (if an arbitrarily selected object has G if it has F, then everything with F has G). With the adoption of new letters ' a ', ' b ', ' c ' goes a natural extension of UE: from $(x)(Fx \rightarrow Gx)$, for example, we can conclude not only that $Fm \rightarrow Gm$ but also that $Fa \rightarrow Ga$, $Fb \rightarrow Gb$, and so on (arbitrarily selected objects from the universe are after all particular objects in the universe, so that what holds of everything holds of them too). The rule UI and this extension of UE are both illustrated in the following proofs:

102 $(x)(Fx \rightarrow Gx)$, $(x)(Gx \rightarrow Hx) \vdash (x)(Fx \rightarrow Hx)$

1	(1)	$(x)(Fx \rightarrow Gx)$	A
2	(2)	$(x)(Gx \rightarrow Hx)$	A
1	(3)	$Fa \rightarrow Ga$	1 UE
2	(4)	$Ga \rightarrow Ha$	2 UE
1,2	(5)	$Fa \rightarrow Ha$	3,4 SI(S) 1.2.1(i)
1,2	(6)	$(x)(Fx \rightarrow Hx)$	5 UI

To prove that $(x)(Fx \rightarrow Hx)$, we aim to prove $Fa \rightarrow Ha$ (to prove that everything with F has H we aim to prove that an arbitrarily selected object with F has H). From assumptions (1) and (2) by UE in its newly extended form we have (3) $Fa \rightarrow Ga$ and (4) $Ga \rightarrow Ha$; the desired $Fa \rightarrow Ga$ now follows by propositional calculus reasoning, steps embodied in a sequent from Chapter 1, which we here abbreviate by SI. (Strictly, we have not proved that SI is obtainable as a derived rule for the predicate calculus, but the extension of our demonstration in Chapter 2, Section 2, to the new formal language is in fact immediate.) This proof is typical of predicate calculus work where both assumptions and conclusions are universally quantified: we drop the universal quantifiers from assumptions, changing variables to arbitrary names, apply *propositional calculus* steps, and finally reintroduce a universal quantifier by UI. Here is another example.

103 $(x)(Fx \rightarrow Gx), (x)Fx \vdash (x)Gx$

1	(1)	$(x)(Fx \rightarrow Gx)$	A
2	(2)	$(x)Fx$	A
1	(3)	$Fa \rightarrow Ga$	1 UE
2	(4)	Fa	2 UE
1,2	(5)	Ga	3,4 MPP
1,2	(6)	$(x)Gx$	5 UI

To prove $(x)Gx$ by UI we aim for Ga, which follows by MPP from $Fa \rightarrow Ga$ and Fa, obtainable from the assumptions by UE.

As already indicated, some restriction has to be placed on the free use of UI, if fallacies are to be avoided. The following illustration should help to show why. Suppose that, in a geometrical context, we arbitrarily select a shape a and assume (i) that it is acute-angled (that is, that none of its angles are as great as a right angle), and (ii) that it is rectilinear (that is, that it is formed by straight lines); then by elementary geometrical reasoning we can conclude that a is a triangle. Expressing (i) by 'Aa', (ii) by 'Ra', and the conclusion by 'Ta', we have that Ta follows from Aa and Ra. Hence, by a step of CP, given that Aa, $Ra \rightarrow Ta$. If we now apply UI as it stands, from Aa we can conclude that $(x)(Rx \rightarrow Tx)$—given an arbitrarily selected acute-angled shape, then all rectilinear shapes are triangles. The conclusion is evidently false, yet we can make the assumption true by simply selecting an acute-angled shape.

The fallacy involved here may be described by saying that we have no right to pass from the conclusion $Ra \rightarrow Ta$ to $(x)(Rx \rightarrow Tx)$, just because that conclusion rests on the *special assumption* concerning *a* that *Aa*. We have in fact proved that if our arbitrarily selected shape *a* is rectilinear then it is triangular, but only on the assumption that it is acute-angled as well. We can avoid this fallacy if, before we apply UI in passing from a proposition about *a* to a universal conclusion, we make sure that the assumptions on which the proposition about *a* rest do not include a special assumption concerning *a* itself; that is to say that, before we apply UI, we should make sure that ' *a* ' does not appear in any of the assumptions on which the conclusion rests. This blocks successfully the fallacious move indicated above. For the conclusion $Ra \rightarrow Ta$ rested on the assumption *Aa*, in which *a* is mentioned, so that UI cannot be applied.

The applications of UI given earlier obey this restriction, as the reader should check for himself. For example, in the proof of 103 we applied UI to the conclusion *Ga* to obtain $(x)Gx$; but the assumptions on which *Ga* rested were $(x)(Fx \rightarrow Gx)$ and $(x)Fx$, in neither of which does ' *a* ' appear. The restriction is easy to observe in practice: before applying UI to ' $\ldots a \ldots$ ', in order to obtain ' $(x)(\ldots x \ldots)$ ', we go through the assumptions on which ' $\ldots a \ldots$ ' rests to ensure that ' *a* ' nowhere appears in them.

The most direct form of the fallacy is observed in the following ' proof ':

| 1 | (1) *Fa* | A |
| 1 | (2) $(x)Fx$ | 1 UI |

For example, taking ' *F* ' as being odd, we may arbitrarily select, in the universe of numbers, an odd number, say 3, so that *Fa* becomes true; but it evidently does not follow that *all* numbers are odd, which is false. The move from (1) to (2) is prevented by the restriction, since (1) depends on *itself*, in which ' *a* ' appears.

We have not, in fact, in this section given precise formulations of the rules UE and UI; this is delayed until Chapter 4, Section 1, where we present detailed formation rules for the predicate calculus analogous to those in Chapter 2, Section 1, for the propositional calculus. But the present intuitive account should enable the student to understand the elementary proofs given in the text and to work the exercises that follow. It is only in more sophisticated work that we require an exact statement of the quantifier rules.

EXERCISES

1 Translate the following arguments into the symbolism of the predicate calculus, and then show their validity by UE and propositional calculus rules:

   (*a*) Jacques is a Frenchman; all Frenchmen are niggardly: therefore Jacques is niggardly. ('*m*', '*F*', '*N*')

   (*b*) Jacques is niggardly; no Frenchmen are niggardly; therefore Jacques is not a Frenchman. ('*m*', '*N*', '*F*')

   (*c*) William is not a Frenchman; only Frenchmen are avaricious; therefore William is not avaricious. ('*n*', '*F*', '*A*')

   (*d*) All male nurses are sympahetic; William is not sympathetic; William is male; therefore William is not a nurse. ('*M*', '*N*', '*S*', '*n*')

   (*e*) All Frenchmen except Parisians are kindly; Jacques is a Frenchman; Jacques is not kindly; therefore Jacques is a Parisian. ('*F*', '*P*', '*K*', '*m*')

2 (i) Using UE and UI together with propositional calculus rules, show the validity of the following sequents:

   (*a*) $(x)(Fx \rightarrow Gx), (x)(Gx \rightarrow -Hx) \vdash (x)(Fx \rightarrow -Hx)$

   (*b*) $(x)(Fx \rightarrow -Gx), (x)(Hx \rightarrow Gx) \vdash (x)(Fx \rightarrow -Hx)$

   (*c*) $(x)(Fx \rightarrow Gx), (x)(Hx \rightarrow -Gx) \vdash (x)(Fx \rightarrow -Hx)$

   (*d*) $(x)(Gx \rightarrow -Fx), (x)(Hx \rightarrow Gx) \vdash (x)(Fx \rightarrow -Hx)$

   (*e*) $(x)(Fx \rightarrow Gx) \vdash (x) Fx \rightarrow (x) Gx$

   (*f*) $(x)(Fx \lor Gx \rightarrow Hx), (x)-Hx \vdash (x)-Fx$

   (ii) For each of the following arguments, indicate which of the sequents (*a*)–(*d*) above exhibits its logical form (thus establishing the validity of the arguments):

   (*a*) No Germans are Frenchmen; all Huns are German; therefore no Frenchmen are Huns.

   (*b*) No Frenchmen are fanatics; all Huns are fanatics; therefore no Frenchmen are Huns.

   (*c*) All Huns are fanatics; no Frenchmen are fanatics; therefore no Huns are Frenchmen.

   (*d*) All Germans are fanatics; no fanatics are histrionic; therefore no Germans are histrionic.

3 THE EXISTENTIAL QUANTIFIER

As the universal quantifier is related to ' & ', so is the existential quantifier to ' v '. In the universe of 3 objects discussed in the last section, ' (x) Fx ' meant the same as ' Fm & Fn & Fo '. Now to say that there is *at least one* x with F in this universe is to say that *either m* has F or n has F or o has F. Hence here ' (∃x)Fx ' means the same as ' Fm v Fn v Fo '. In the case of an infinitely large universe, say that of the natural numbers, to say that there is a number with F or that some number has F is to say that either 0 has F or 1 has F or 2 has F or . . . . As we need the universal quantifier because we cannot write down an ' infinite conjunction ', so we need the existential quantifier because we cannot write down an ' infinite disjunction '.

Accordingly, the two rules for the existential quantifier can be seen as extensions of the rules vI and vE. Let us take the rule of existential quantifier introduction (EI) first. To establish a conclusion such as (∃x)Fx, a natural premiss is something like Fm: given a particular object with F, we can conclude that *something* has F. Thus, in our universe of 3 objects, given any one of Fm, Fn, Fo, we can conclude (∃x)Fx; or, in the infinite case, given any particular natural number with F, we can conclude that some number has F. If we bear in mind the disjunctive status of the existential quantifier, the analogy with vI should be obvious.

Hence the justification for EI is that, if a particular thing has a certain property, then something must have it, and EI will enable us to pass from premisses such as Fm and Fn to conclusion (∃x)Fx, and from Fm & Gm and Fn & Gn to conclusion (∃x)(Fx & Gx) (if m has both F and G, or if n has both F and G, then something has both F and G). Further, we extend the rule to apply also to premisses concerning arbitrarily selected objects a, b, c: for, if an arbitrarily selected thing has F, then again something has F. Hence, for example, EI will enable us to pass from premiss Fa v Ga to conclusion (∃x)(Fx v Gx) (if an arbitrarily selected object has either F or G, then something has either F or G).

A very simple application of this rule occurs in the proof of the following (evidently valid) sequent:

104 $(x)Fx \vdash (\exists x)Fx$

1	(1)	$(x)Fx$	A
1	(2)	Fa	1 UE
1	(3)	$(\exists x)Fx$	2 EI

If everything has F, then in particular an arbitrarily selected object a has F, whence by EI something has F.

The rule of existential quantifier elimination (EE) can best be understood in the light of the rule vE. Given a disjunction A v B, it being desired to establish a conclusion C, we derive C first from A as assumption and then from B as assumption, knowing that, if C follows from *both* A *and* B, then, since one or the other holds, C must hold. Similarly, if we know in our 3-object universe that something has F, we know effectively

(1) *Fm* v *Fn* v *Fo*.

Seeking to establish a conclusion C, we might assume each disjunct of the complex disjunction in turn, knowing that if C follows from *all* those disjuncts, then, since one or other holds, C must hold. However, where an infinite universe is involved, $(\exists x)Fx$ is a kind of 'infinite disjunction', and there can be no question of deriving C from each of the infinitely many disjuncts. Now in the case of UI, we adopted the device of arbitrary names 'a', 'b', 'c' just because we could not establish separately the infinitely many conjuncts that go to make the 'infinite conjunction' $(x)Fx$. For EE we may use the same device. Instead of showing that C follows from the separate assumptions Fm, Fn, Fo, we may show instead that C follows from the single assumption, Fa, that an arbitrarily selected object has F. The pattern of proof will then be: given $(\exists x)Fx$, and that C follows from assumption Fa, then C follows anyway. Here the proof of C from Fa is a condensed representation of possibly infinitely many derivations of C from all the disjuncts in the disguised disjunction $(\exists x)Fx$. We may call Fa here, I hope suggestively, the *typical disjunct* corresponding to the existential proposition $(\exists x)Fx$.

Thus the justification for EE is somewhat as follows. If something has a certain property, and if it can be shown that a conclusion C follows from the assumption that an arbitrarily selected object has that property, then we know that C holds; for if something has the property, and no matter which has it then C holds, then C holds anyway. The conclusion C will of course, as in vE, rest on any assumptions on which the existential proposition rests, and on any assumptions used to derive C from the corresponding typical disjunct apart from the disjunct itself. And on the right-hand side we shall cite three lines: (i) the line where the existential proposition occurs; (ii) the line where the typical disjunct is assumed; and (iii) the line where C is drawn as conclusion from the typical disjunct as assumption.

These new rules are illustrated by the following proofs:

105 $(x)(Fx \rightarrow Gx), (\exists x)Fx \vdash (\exists x)Gx$

1	(1)	$(x)(Fx \rightarrow Gx)$	A
2	(2)	$(\exists x)Fx$	A
3	(3)	Fa	A
1	(4)	$Fa \rightarrow Ga$	1 UE
1,3	(5)	Ga	3,4 MPP
1,3	(6)	$(\exists x)Gx$	5 EI
1,2	(7)	$(\exists x)Gx$	2,3,6 EE

Given that everything with F has G and that something has F, we show that something has G. We assume, preparatory to EE, that an arbitrarily selected object a has F at line (3), and then conclude (line (6)) that something has G. We are now ready for a step of EE; given an existential proposition to the effect that something has F at line (2) and a derivation of the desired conclusion from the corresponding typical disjunct at line (6), we obtain the conclusion again at line (7). We cite on the right line (2), the existential proposition, line (3), the typical disjunct, and line (6), the conclusion obtained from that assumption. The conclusion now rests upon whatever assumptions the existential proposition rests upon—here merely itself—and any assumptions used to obtain the conclusion from the typical disjunct Fa apart from Fa itself, which gives just (1) and (2).

The analogy with vE can be brought out by supposing that, as a special case, we are dealing with a 2-object universe, containing just m and n. Then, for this universe, $(\exists x)Fx$ amounts to $Fm \vee Fn$, and $(\exists x)Gx$ to $Gm \vee Gn$. The corresponding proof with vE in place of EE would go as follows:

	1	(1) $(x)(Fx \rightarrow Gx)$	A			
	2	(2) $Fm \lor Fn$	A			
3	(3) Fm	A		3′	(3′) Fn	A
1	(4) $Fm \rightarrow Gm$	1 UE		1	(4′) $Fn \rightarrow Gn$	1 UE
1,3	(5) Gm	3, 4 MPP		1,3′	(5′) Gn	3′, 4′ MPP
1,3	(6) $Gm \lor Gn$	5 vI		1,3′	(6′) $Gm \lor Gn$	5′ vI
	1,2	(7) $Gm \lor Gn$			2,3,6,3′,6′ vE	

Here the lines (3′)–(5′) exactly mirror (3)–(5) with ' *n* ' in place of ' *m* '. The lines (3)–(6) of our actual proof of 105 condense these twin arguments into one argument, by the employment of arbitrary names in place of the proper names ' *m* ' and ' *n* ', and by using the typical disjunct ' *Fa* ' in place of the separate disjuncts ' *Fm* ' and ' *Fn* '.

106 $(x)(Gx \rightarrow Hx), (\exists x)(Fx \,\&\, Gx) \vdash (\exists x)(Fx \,\&\, Hx)$

1	(1) $(x)(Gx \rightarrow Hx)$	A
2	(2) $(\exists x)(Fx \,\&\, Gx)$	A
3	(3) $Fa \,\&\, Ga$	A
1	(4) $Ga \rightarrow Ha$	1 UE
3	(5) Ga	3 &E
1,3	(6) Ha	4,5 MPP
3	(7) Fa	3 &E
1,3	(8) $Fa \,\&\, Ha$	6,7 &I
1,3	(9) $(\exists x)(Fx \,\&\, Hx)$	8 EI
1,2	(10) $(\exists x)(Fx \,\&\, Hx)$	2,3,9 EE

The strategy here should be clear. To prove $(\exists x)(Fx \,\&\, Hx)$ from $(\exists x)(Fx \,\&\, Gx)$, we aim for the same conclusion from $Fa \,\&\, Ga$, the corresponding typical disjunct. Since everything with G has H, from Ga we can infer Ha, hence a has both F and H, and so something has both F and H. The conclusion at (10) rests on (2), the original existential proposition, and (1), which was used to obtain the conclusion from (3), as we see at line (9).

These two proofs illustrate a general tip for proof-discovery. Given $(\exists x)(\ldots x \ldots)$ and desiring to prove a conclusion C, you should assume (. . . *a* . . .) as typical disjunct and try to obtain C from it. For, if you succeed, EE will give you just this conclusion. Once (. . . *a* . . .) has been assumed, reasoning of the propositional calculus type will generally assist in the derivation of C.

As in the case of UI, the use of arbitrary names with EE necessitates certain restrictions if fallacies are to be avoided. In the case of UI, we required that the arbitrary name in question should not appear in the assumptions on which the conclusion drawn rested. For EE we require that the arbitrary name in question shall not appear either in the conclusion C drawn or in the assumptions used to derive C from the typical disjunct (though of course it will appear in the typical disjunct itself).

To see that the arbitrary name must not appear in the conclusion C, we need only observe that otherwise we could prove, given that something has F, that everything has F.

1	(1) $(\exists x)Fx$	A
2	(2) Fa	A
1	(3) Fa	1,2,2 EE
1	(4) $(x)Fx$	3 UI

The step of UI is correct, since 1 does not contain '*a*'. But the step of EE is incorrect because the conclusion in question, here Fa, does contain '*a*'. It does not follow from something's having F that an arbitrarily selected object has F, though of course Fa follows from itself. To see that the arbitrary name must not appear in the assumptions (apart from the typical disjunct) used to obtain C, consider the following 'proof':

1	(1) Fa	A
2	(2) $(\exists x)Gx$	A
3	(3) Ga	A
1,3	(4) $Fa \& Ga$	1,3 &I
1,3	(5) $(\exists x)(Fx \& Gx)$	4 EI
1,2	(6) $(\exists x)(Fx \& Gx)$	2,3,5 EE

The conclusion, that something has both F and G, is here reached from the two assumptions that an arbitrarily selected object has F and that something has G. Now, let F be being even, and G be being

115

odd: then I can select a number a which is even, so that Fa becomes true, and there are odd numbers, so that $(\exists x)Gx$ is also true. But it is false that any number is both odd and even. The step of EE is unsound, because the conclusion at line (5) rests on (1) which contains 'a'.

The new restriction is again easy to observe in practice. For example, to see that the step of EE at line (10) of 106 is correct, we inspect line (9); the conclusion there does not contain 'a': of the two assumptions on which it rests, (3), the typical disjunct, of course contains 'a' but (1) does not; thus the restriction is met.

Since arbitrarily selected objects play a large part in our work, it may be as well to attempt to clarify their position. Let F be some property, and a an arbitrarily selected object from some universe; then, given that everything has F, a has F, but not conversely. We accept as valid the sequent $(x)Fx \vdash Fa$, but not the sequent $Fa \vdash (x)Fx$; and we reject the latter because a, though arbitrarily selected, may not be *typical*. On the other hand, by UI, under certain conditions we pass from premiss $(\ldots a \ldots)$ to conclusion $(x)(\ldots x \ldots)$; however, the conditions involved are such as to ensure that a *is* here typical, for we stipulate that $(\ldots a \ldots)$ shall not rest on any special assumptions about a. We also declare that, given that a has F, something has F, but not conversely. We accept as valid the sequent $Fa \vdash (\exists x)Fx$ but not the sequent $(\exists x)Fx \vdash Fa$, and we reject the latter because a, being arbitrarily selected, may not be one of the given objects with F. On the other hand, by EE, under certain conditions we can derive conclusions obtained from Fa directly from $(\exists x)Fx$, as though what Fa implied $(\exists x)Fx$ implied also; however, the conditions involved are such as to ensure that any such conclusion is obtained from Fa only on the understanding that a is typical—no special assumptions about a other than Fa are made and the conclusion does not concern a—and so can be taken as one of the given objects with F. Thus the claim that an arbitrarily selected object has F must be distinguished both from the claim $(x)Fx$ and the claim $(\exists x)Fx$, though it is derivable from the former and the latter is derivable from it.

EXERCISES

1 Using quantifier and propositional calculus rules, show the validity of the following sequents:

    (*a*) $(x)(Fx \rightarrow Gx)$, $(\exists x) -Gx \vdash (\exists x) -Fx$

    (*b*) $(x)(Fx \rightarrow Gx \,\&\, Hx)$, $(\exists x)Fx \vdash (\exists x)Hx$

    (*c*) $(x)(Fx \vee Gx \rightarrow Hx)$, $(\exists x) -Hx \vdash (\exists x) -Fx$

2   (i) Using quantifier and propositional calculus rules, show the validity of the following sequents:

    (*a*) $(x)(Gx \rightarrow -Hx)$, $(\exists x)(Fx \,\&\, Gx) \vdash (\exists x)(Fx \,\&\, -Hx)$

    (*b*) $(x)(Hx \rightarrow Gx)$, $(\exists x)(Fx \,\&\, -Gx) \vdash (\exists x)(Fx \,\&\, -Hx)$

    (*c*) $(x)(Hx \rightarrow -Gx)$, $(\exists x)(Fx \,\&\, Gx) \vdash (\exists x)(Fx \,\&\, -Hx)$

    (*d*) $(x)(Gx \rightarrow Hx)$, $(\exists x)(Gx \,\&\, Fx) \vdash (\exists x)(Fx \,\&\, Hx)$

    (*e*) $(\exists x)(Gx \,\&\, Hx)$, $(x)(Gx \rightarrow Fx) \vdash (\exists x)(Fx \,\&\, Hx)$

    (*f*) $(x)(Gx \rightarrow -Hx)$, $(\exists x)(Gx \,\&\, Fx) \vdash (\exists x)(Fx \,\&\, -Hx)$

    (*g*) $(\exists x)(Gx \,\&\, -Hx)$, $(x)(Gx \rightarrow Fx) \vdash (\exists x)(Fx \,\&\, -Hx)$

  (ii) For each of the following arguments, indicate which of the sequents (*a*)–(*g*) above exhibits its logical form (thus establishing the validity of the arguments):

    (*a*) No mountains are climbable; some hills are climbable; therefore some hills are not mountains.

    (*b*) Some mountains are climbable; all mountains are hills; therefore some hills are climbable.

    (*c*) All hills are climbable; some mountains are not climbable; therefore some mountains are not hills.

    (*d*) All hills are climbable; some hills are mountains; therefore some mountains are climbable.

    (*e*) No mountains are climbable; some hills are mountains; therefore some hills are not climbable.

    (*f*) Some mountains are not climbable; all mountains are hills; therefore some hills are not climbable.

    (*g*) No mountains are climbable; some mountains are hills; therefore some hills are not climbable.

4 ELEMENTARY VALID SEQUENTS WITH QUANTIFIERS

In sequents 100–106, we have observed some of the basic results concerning quantifiers which we are now in a position to establish. This section is devoted to proving further such results; these are important in themselves, and their proofs will incidentally afford greater insight into the use of the quantifier rules.

107 $(x)(Fx \to Gx) \vdash (x)Fx \to (x)Gx$

108 $(x)(Fx \to Gx) \vdash (\exists x)Fx \to (\exists x)Gx$

Given that everything with F has G, it follows that if everything has F everything has G (107), and it follows that if something has F something has G (108). The proofs are immediate from the proofs of 103 and 105, by a further step of CP in each case.

109 $(x)(Fx \And Gx) \dashv\vdash (x)Fx \And (x)Gx$

(a) $(x)(Fx \And Gx) \vdash (x)Fx \And (x)Gx$

1	(1)	$(x)(Fx \And Gx)$	A
1	(2)	$Fa \And Ga$	1 UE
1	(3)	Fa	2 &E
1	(4)	$(x)Fx$	3 UI
1	(5)	Ga	2 &E
1	(6)	$(x)Gx$	5 UI
1	(7)	$(x)Fx \And (x)Gx$	4,6 &I

(b) $(x)Fx \And (x)Gx \vdash (x)(Fx \And Gx)$

1	(1)	$(x)Fx \And (x)Gx$	A
1	(2)	$(x)Fx$	1 &E
1	(3)	Fa	2 UE
1	(4)	$(x)Gx$	1 &E
1	(5)	Ga	4 UE
1	(6)	$Fa \And Ga$	3,5 &I
1	(7)	$(x)(Fx \And Gx)$	6 UI

The proposition that everything has both F and G is interderivable with the proposition that both everything has F and everything has G. The proofs require little comment, except to observe that the restriction on UI is met at lines (4) and (6) of (a) and at line (7) of (b), since ' a ' does not appear in assumption (1) of either proof.

110 $(\exists x)(Fx \lor Gx) \dashv\vdash (\exists x)Fx \lor (\exists x)Gx$

(a) $(\exists x)(Fx \lor Gx) \vdash (\exists x)Fx \lor (\exists x)Gx$

1	(1)	$(\exists x)(Fx \lor Gx)$	A

2	(2) *Fa* v *Ga*	A
3	(3) *Fa*	A
3	(4) (∃x)*Fx*	3 EI
3	(5) (∃x)*Fx* v (∃x)*Gx*	4 vI
6	(6) *Ga*	A
6	(7) (∃x)*Gx*	6 EI
6	(8) (∃x)*Fx* v (∃x)*Gx*	7 vI
2	(9) (∃x)*Fx* v (∃x)*Gx*	2,3,5,6,8 vE
1	(10) (∃x)*Fx* v (∃x)*Gx*	1,2,9 EE

(*b*) (∃x)*Fx* v (∃x)*Gx* ⊢ (∃x)(*Fx* v *Gx*)

1	(1) (∃x)*Fx* v (∃x)*Gx*	A
2	(2) (∃x)*Fx*	A
3	(3) *Fa*	A
3	(4) *Fa* v *Ga*	3 vI
3	(5) (∃x)(*Fx* v *Gx*)	4 EI
2	(6) (∃x)(*Fx* v *Gx*)	2,3,5 EE
7	(7) (∃x)*Gx*	A
8	(8) *Ga*	A
8	(9) *Fa* v *Ga*	8 vI
8	(10) (∃x)(*Fx* v *Gx*)	9 EI
7	(11) (∃x)(*Fx* v *Gx*)	7,8,10 EE
1	(12) (∃x)(*Fx* v *Gx*)	1,2,6,7,11 vE

The proposition that something has either *F* or *G* is interderivable with the proposition that either something has *F* or something has *G*. In proof (*a*), the overall strategy is, given an existential proposition as assumption at line (1), to assume its corresponding typical disjunct (line (2)) and to obtain the desired conclusion from that. This is achieved at line (9) by vE; vE is involved since the typical disjunct is itself a disjunction, so that we obtain the conclusion from each limb in turn (lines (5) and (8)). At the final step of EE, we notice that the conclusion does not contain ' *a* ' and that line (9) rests only on (2), the typical disjunct itself; hence (10) rests only on (1), and the restriction on EE is met. In proof (*b*), our overall strategy is to proceed by vE, given a disjunction as assumption at line (1). Each disjunct being an existential proposition, after assuming it (lines (2) and (7)) we assume its corresponding typical disjunct (lines (3) and (8)) and obtain the conclusion from that (lines (5) and (10)). For control of the two EE steps (lines (6) and (11)), we notice that the conclusion lacks ' *a* ' and that lines (5) and (10) rest only on the typical disjuncts (3) and (8).

The interderivability results 109 and 110 are entirely to be expected when we bear in mind the conjunctive status of the universal quantifier and the disjunctive status of the existential quantifier, as discussed in the previous two sections. Put loosely, they claim that a universal quantifier may be distributed through a conjunction and an existential quantifier through a disjunction.

111 $(\exists x)(Fx \mathbin{\&} Gx) \vdash (\exists x)Fx \mathbin{\&} (\exists x)Gx$

1	(1) $(\exists x)(Fx \mathbin{\&} Gx)$	A
2	(2) $Fa \mathbin{\&} Ga$	A
2	(3) Fa	2 &E
2	(4) $(\exists x)Fx$	3 EI
2	(5) Ga	2 &E
2	(6) $(\exists x)Gx$	5 EI
2	(7) $(\exists x)Fx \mathbin{\&} (\exists x)Gx$	4,6 &I
1	(8) $(\exists x)Fx \mathbin{\&} (\exists x)Gx$	1,2,7 EE

Given that something has both F and G, it follows that something has F and something has G. We proceed by EE, and assume at line (2) the typical disjunct $Fa \mathbin{\&} Ga$ corresponding to (1) $(\exists x)$ $(Fx \mathbin{\&} Gx)$. For EE at line (8), we observe that the conclusion lacks ' *a* ' and that (7) rests only on (2).

The converse sequent, $(\exists x)Fx \mathbin{\&} (\exists x)Gx \vdash (\exists x)(Fx \mathbin{\&} Gx)$, is *not* valid; consider the universe of positive integers, and let F be the property of being even, G the property of being odd; then it is true that there are even numbers and that there are odd numbers $((\exists x)Fx$ $\mathbin{\&} (\exists x)Gx)$, but false that there are numbers both even and odd $((\exists x)(Fx \mathbin{\&} Gx))$. It is instructive to see how natural attempts to prove this sequent fail, in view of the restriction on EE. We might start:

1	(1) $(\exists x)Fx$ & $(\exists x)Gx$	A
1	(2) $(\exists x)Fx$	1 &E
1	(3) $(\exists x)Gx$	1 &E
4	(4) Fa	A
5	(5) Ga	A
4,5	(6) Fa & Ga	4,5 &I
4,5	(7) $(\exists x)(Fx$ & $Gx)$	6 EI

For the existential propositions (2) and (3) we have assumed the typical disjuncts (4) and (5), and derived the conclusion $(\exists x)(Fx$ & $Gx)$ from them. But any attempt to apply EE, either using (2) or using (3), now fails, since the conclusion at line (7) rests on (4) and (5), in *both* of which 'a' appears. Hence we can obtain neither

1,5	(8) $(\exists x)(Fx$ & $Gx)$	2,4,7 EE

(since ' a ' appears in (5)), nor

1,4	(8) $(\exists x)(Fx$ & $Gx)$	3,5,7 EE

(since ' a ' appears in (4)). If we could reach either of these lines, the conclusion

1	(9) $(\exists x)(Fx$ & $Gx)$	

would of course follow by a further (sound) step of EE.

112 $(x)Fx \lor (x)Gx \vdash (x)(Fx \lor Gx)$

1	(1) $(x)Fx \lor (x)Gx$	A
2	(2) $(x)Fx$	A
2	(3) Fa	2 UE
2	(4) $Fa \lor Ga$	3 vI
2	(5) $(x)(Fx \lor Gx)$	4 UI
6	(6) $(x)Gx$	A
6	(7) Ga	6 UE
6	(8) $Fa \lor Ga$	7 vI
6	(9) $(x)(Fx \lor Gx)$	8 UI
1	(10) $(x)(Fx \lor Gx)$	1,2,5,6,9 vE

Given that either everything has F or everything has G, then everything has either F or G. Proof is by vE; in using UI at lines (5) and (9), we observe that neither limb of the assumed disjunction (1) contains ' a ', so that the restriction is met.

The converse sequent, $(x)(Fx \vee Gx) \vdash (x)Fx \vee (x)Gx$, is *not* valid, and the interpretation just used shows why not; for all positive integers are either even or odd, but it is neither the case that all are even nor the case that all are odd. In this case, it is the restriction on UI that prevents natural attempts to prove the sequent. Thus:

1	(1) $(x)(Fx \vee Gx)$	A
1	(2) $Fa \vee Ga$	1 UE
3	(3) Fa	A

Concluding $Fa \vee Ga$ from (1), we assume the first disjunct Fa at line (3); but now we are prevented from concluding $(x) Fx$ since (3) contains ' a '. If this step were permitted, we could conclude $(x)Fx \vee (x)Gx$ by vI, then obtain the same conclusion from Ga, and vE would yield the invalid sequent.

113 $(\exists x)Fx \dashv\vdash -(x)-Fx$

    (a) $(\exists x)Fx \vdash -(x)-Fx$

1	(1) $(\exists x)Fx$	A
2	(2) Fa	A
3	(3) $(x)-Fx$	A
3	(4) $-Fa$	3 UE
2,3	(5) $Fa \ \& \ -Fa$	2,4 &I
2	(6) $--(x)-Fx$	3,5 RAA
1	(7) $-(x)- Fx$	1,2,6 EE

    (b) $-(x)-Fx \vdash (\exists x)Fx$

1	(1) $-(x)-Fx$	A
2	(2) $-(\exists x)Fx$	A
3	(3) Fa	A
3	(4) $(\exists x)Fx$	3 EI
2,3	(5) $(\exists x)Fx \ \& \ -(\exists x)Fx$	2,4 &I

2	(6) $-Fa$	3,5 RAA
2	(7) $(x)-Fx$	6 UI
1,2	(8) $(x)-Fx$ & $-(x)-Fx$	1,8 &I
1	(9) $--(\exists x)Fx$	2,8 RAA
1	(10) $(\exists x)Fx$	9 DN

The proposition that something has F is interderivable with the proposition that it is not the case that everything lacks F. In proof (*a*), given $(\exists x)Fx$, we assume the typical disjunct Fa (line (2)) and aim for the conclusion $-(x)-Fx$ from that. We obtain this by RAA, and assume $(x)-Fx$ at line (3) accordingly. The restriction on EE at line (7) is met, since the conclusion lacks ' a '. In proof (*b*), we assume at line (2) $-(\exists x)Fx$, and aim to derive $(x)-Fx$, contradicting (1). To obtain $(x)-Fx$, it suffices to obtain $-Fa$ and use UI, so we assume Fa (line (3)) and go for a contradiction (line (5)). In applying UI at line (7), note that (6) rests only on (2), which lacks ' a '.

By the relationship between the quantifiers and ' & ' and ' v ', the interderivability result 113 is akin to 36, in Chapter 1, Section 5. Indeed, *mutatis mutandis*, the proofs are structurally the same, as the reader may care to check for himself. Put loosely, 113 tells us that any existential proposition is tantamount to the negation of a universal proposition, in the way in which 36 tells us that any disjunction is tantamount to the negation of a certain conjunction. Our next result, conversely, says that any universal proposition is tantamount to the negation of an existential proposition, and should be compared with the sequent 1.5.1(*h*).

114 $(x)Fx \dashv\vdash -(\exists x)-Fx$

(*a*) $(x)Fx \vdash -(\exists x)-Fx$

1	(1) $(x)Fx$	A
2	(2) $(\exists x)-Fx$	A
3	(3) $-Fa$	A
1	(4) Fa	1 UE
1,3	(5) Fa & $-Fa$	3,4 &I
3	(6) $-(x)Fx$	1,5 RAA

2	(7) $-(x)Fx$	2,3,6 EE
1,2	(8) $(x)Fx$ & $-(x)Fx$	1,7 &I
1	(9) $-(\exists x)-Fx$	2,8 RAA

(b) $-(\exists x)-Fx \vdash (x)Fx$

1	(1) $-(\exists x)-Fx$	A
2	(2) $-Fa$	A
2	(3) $(\exists x)-Fx$	2 EI
1,2	(4) $(\exists x)-Fx$ & $-(\exists x)-Fx$	1,3 &I
1	(5) $--Fa$	2,4 RAA
1	(6) Fa	5 DN
1	(7) $(x)Fx$	6 UI

The proposition that everything has F is interderivable with the proposition that it is not the case that something lacks F. In proof (a), we assume $(\exists x)-Fx$ (line (2)) to obtain $-(x)Fx$, contradicting (1), and proceed by EE, assuming the typical disjunct at line (3). In proof (b), it suffices to prove Fa from (1), hence we assume $-Fa$ at line (2) and search for a contradiction. The reader should confirm that restrictions on EE and UI are met here.

115 $(x)Fx \dashv\vdash (y)Fy$

1	(1) $(x)Fx$	A
1	(2) Fa	1 UE
1	(3) $(y)Fy$	2 UI

The converse is similarly derivable. Since variables are merely devices for cross-reference, we should expect this interderivability. Both sentences in fact express the same proposition—that everything has F. Similarly:

116 $(\exists x)Fx \dashv\vdash (\exists y)Fy$

1	(1) $(\exists x)Fx$	A
2	(2) Fa	A
2	(3) $(\exists y)Fy$	2 EI
1	(4) $(\exists y)Fy$	1,2,3 EE

The next proof requires no comment; but the reader should watch the observance of the restrictions on EE and UI.

117 $(x)(Fx \rightarrow Gx) \dashv\vdash -(\exists x)(Fx \mathbin{\&} -Gx)$

    (a) $(x)(Fx \rightarrow Gx) \vdash -(\exists x)(Fx \mathbin{\&} -Gx)$

1	(1) $(x)(Fx \rightarrow Gx)$	A
2	(2) $(\exists x)(Fx \mathbin{\&} -Gx)$	A
3	(3) $Fa \mathbin{\&} -Ga$	A
3	(4) $-(Fa \rightarrow Ga)$	3 SI(S) 2.2.5(g)
1	(5) $Fa \rightarrow Ga$	1 UE
1,3	(6) $(Fa \rightarrow Ga) \mathbin{\&}$ $-(Fa \rightarrow Ga)$	4,5 &I
3	(7) $-(x)(Fx \rightarrow Gx)$	1,6 RAA
2	(8) $-(x)(Fx \rightarrow Gx)$	2,3,7 EE
1,2	(9) $(x)(Fx \rightarrow Gx) \mathbin{\&}$ $-(x)(Fx \rightarrow Gx)$	1,8 &I
1	(10) $-(\exists x)(Fx \mathbin{\&} -Gx)$	2,9 RAA

    (b) $- (\exists x)(Fx \mathbin{\&} -Gx) \vdash (x)(Fx \rightarrow Gx)$

1	(1) $-(\exists x)(Fx \mathbin{\&} -Gx)$	A
2	(2) $-(Fa \rightarrow Ga)$	A
2	(3) $Fa \mathbin{\&} -Ga$	2 SI(S) 2.2.5(g)
2	(4) $(\exists x)(Fx \mathbin{\&} -Gx)$	3 EI
1,2	(5) $(\exists x)(Fx \mathbin{\&} -Gx) \mathbin{\&}$ $-(\exists x)(Fx \mathbin{\&} -Gx)$	1,4 &I
1	(6) $--(Fa \rightarrow Ga)$	2,5 RAA
1	(7) $Fa \rightarrow Ga$	6 DN
1	(8) $(x)(Fx \rightarrow Gx)$	7 UI

The proposition that everything with F has G is interderivable with the proposition that it is not the case that something has F but not G; more loosely, to affirm that everything with F has G is to deny that something with F lacks G.

The last two interderivability results of this section are somewhat surprising.

118 $(x)(Fx \to P) \dashv\vdash (\exists x)Fx \to P$

    (a) $(x)(Fx \to P) \vdash (\exists x)Fx \to P$

1	' (1)	$(x)(Fx \to P)$	A
2	(2)	$(\exists x)Fx$	A
3	(3)	Fa	A
1	(4)	$Fa \to P$	1 UE
1,3	(5)	P	3,4 MPP
1,2	(6)	P	2,3,5 EE
1	(7)	$(\exists x)Fx \to P$	2,6 CP

    (b) $(\exists x)Fx \to P \vdash (x)(Fx \to P)$

1	(1)	$(\exists x)Fx \to P$	A
2	(2)	Fa	A
2	(3)	$(\exists x)Fx$	2 EI
1,2	(4)	P	1,3 MPP
1	(5)	$Fa \to P$	2,4 CP
1	(6)	$(x)(Fx \to P)$	5 UI

The universal proposition that, for any object, if it has F then P is interderivable with the conditional that if something has F then P. (It is important to see here that the universal quantifier in '$(x)(Fx \to P)$' governs the whole expression '$(Fx \to P)$', whilst the existential quantifier in '$(\exists x)Fx \to P$' merely governs the *antecedent* of the whole conditional; compare the difference between '$-(P \to Q)$' and '$-P \to Q$'.) Thus, letting F be the property of being a man, and P the proposition that the earth is populated, to say that if there are men then the earth is populated is to say that, for any object, if it is a man then the earth is populated.

119 $(\exists x)(P \to Fx) \dashv\vdash P \to (\exists x)Fx$

    (a) $(\exists x)(P \to Fx) \vdash P \to (\exists x)Fx$

1	(1)	$(\exists x)(P \to Fx)$	A
2	(2)	P	A

3	(3) $P \rightarrow Fa$	A
2,3	(4) Fa	2,3 MPP
2,3	(5) $(\exists x)Fx$	4 EI
1,2	(6) $(\exists x)Fx$	1,3,5 EE
1	(7) $P \rightarrow (\exists x)Fx$	2,6 CP

(b) $P \rightarrow (\exists x)Fx \vdash (\exists x)(P \rightarrow Fx)$

1	(1) $P \rightarrow (\exists x)Fx$	A
	(2) $P \vee - P$	TI 44
3	(3) P	A
1,3	(4) $(\exists x)Fx$	1,3 MPP
5	(5) Fa	A
5	(6) $P \rightarrow Fa$	5 SI(S) 50
5	(7) $(\exists x)(P \rightarrow Fx)$	6 EI
1,3	(8) $(\exists x)(P \rightarrow Fx)$	4,5,7 EE
9	(9) $-P$	A
9	(10) $P \rightarrow Fa$	9 SI(S) 51
9	(11) $(\exists x)(P \rightarrow Fx)$	10 EI
1	(12) $(\exists x)(P \rightarrow Fx)$	2,3,8,9,11 vE

The existential proposition that there is something such that if P then it has F is interderivable with the conditional that if P then something has F. Proof (a) is straightforward, when we note that (3) is the typical disjunct corresponding to (1). Since P at line (2) lacks ' a ', the step of EE at line (6) obeys the restriction. Proof (b) is more complex: it proves convenient to introduce the law of excluded middle (line (2)), and proceed by vE (line (12)). We assume P at line (3), and obtain the conclusion from it at line (8); this phase uses EE, for we reach the conclusion at line (7) from the typical disjunct Fa (line (5)) corresponding to the existential proposition at line (4). The second phase (lines (9)–(11)) uses, like the first phase at line (6), a propositional calculus sequent to pass from $-P$ to the conclusion $P \rightarrow Fa$—a simple substitution-instance on $-P \vdash P \rightarrow Q$. After vE, the conclusion rests only on (1), which was used to obtain it from the first disjunct P at line (8).

127

EXERCISES

1 Establish the following results:

(a) $(x)(Fx \rightarrow Gx) \vdash (x) - Gx \rightarrow (x) - Fx$

(b) $(x)(Fx \rightarrow Gx) \vdash (\exists x) - Gx \rightarrow (\exists x) - Fx$

(c) $(\exists x) - Fx \dashv\vdash -(x)Fx$

(d) $(x) - Fx \dashv\vdash -(\exists x)Fx$

(e) $(x)(Fx \rightarrow -Gx) \dashv\vdash -(\exists x)(Fx \& Gx)$

(f) $(x)(Fx \leftrightarrow Gx) \dashv\vdash (x)(Fx \rightarrow Gx) \& (x)(Gx \rightarrow Fx)$

(g) $(x)(Fx \leftrightarrow Gx) \vdash (x)Fx \leftrightarrow (x) Gx$

(h) $(x)(Fx \leftrightarrow Gx) \vdash (\exists x)Fx \leftrightarrow (\exists x)Gx$

2 (i) Which sequents proved in the text show the interderivability of the proposition that all women are fickle with the proposition that there are not women who are not fickle?

(ii) Which sequents proved in Exercise 1 show the interderivability of the proposition that no men are fickle with the proposition that there are not men who are fickle?

3 Establish the following interderivability results:

(a) $(x)(P \rightarrow Fx) \dashv\vdash P \rightarrow (x)Fx$

(b) $(x)(P \& Fx) \dashv\vdash P \& (x)Fx$

(c) $(\exists x)(P \& Fx) \dashv\vdash P \& (\exists x)Fx$

(d) $(x)(P \lor Fx) \dashv\vdash P \lor (x)Fx$

(e) $(\exists x)(P \lor Fx) \dashv\vdash P \lor (\exists x)Fx$

(f) $(\exists x)(Fx \rightarrow P) \dashv\vdash (x)Fx \rightarrow P$

5 GENERAL QUANTIFIER ARGUMENTS

So far, our derived sequents have concerned properties rather than relations—predicate-letters followed by one variable rather than more than one. We now consider sequents in which relations occur, and the general question of showing the validity of complex arguments as they occur in ordinary speech.

It is convenient to record at the outset two interderivability results which yield *quantifier-shift* principles.

120 $(x)(y)Fxy \dashv\vdash (y)(x)Fxy$

1 (1) $(x)(y)Fxy$ A

1	(2) $(y)Fay$	1 UE
1	(3) Fab	2 UE
1	(4) $(x)Fxb$	3 UI
1	(5) $(y)(x)Fxy$	4 UI

The converse is proved similarly. At line (2), we drop the quantifier ' (x) ' and associated variable ' x ' in favour of an arbitrarily selected object a. Thus (2) says that, for any object, a bears relation F to it. The move from (2) to (3) is similar, and (3) affirms that an arbitrarily selected a bears F to an arbitrarily selected b. Observe that UE permits us to drop only *one* universal quantifier at a time. We then restore the quantifiers in reverse order by UI, noting that neither ' a ' nor ' b ' occurs in (1). From (2), we might have soundly concluded

$\qquad$ 1 $\quad$ (3') Faa

(if a bears F to *every*thing, then a bears F to itself). And from (3') we might have soundly concluded

$\qquad$ 1 $\quad$ (4') $(x)Fxx$

(if arbitrarily chosen a bears F to itself, then under the usual restriction everything bears F to itself). But of course our actual line (4) would not follow from (3'), nor would

$\qquad$ (4″) $(x)Fxa.$

Given that a bears F to itself, it does not follow that everything bears F to a. An arbitrarily selected person has the same age as himself, but not everyone has just that age, though everyone does have just his own age. These considerations should motivate the choice of a different arbitrary name ' b ' at line (3), for otherwise we should not be able to reintroduce two distinct quantifiers.

121 $(\exists x)(\exists y)Fxy \dashv\vdash (\exists y)(\exists x)Fxy$

1	(1) $(\exists x)(\exists y)Fxy$	A
2	(2) $(\exists y)Fay$	A
3	(3) Fab	A
3	(4) $(\exists x)Fxb$	3 EI
3	(5) $(\exists y)(\exists x)Fxy$	4 EI

2   (6) $(\exists y)(\exists x)Fxy$ 2,3,5 EE

1   (7) $(\exists y)(\exists x)Fxy$ 1,2,6 EE

The converse is again proved similarly. (2) is the typical disjunct corresponding to (1), and (3) in turn is the typical disjunct corresponding to (2). We pick distinct arbitrary names ' a ' and ' b ' for good reasons: the step of EE at line (6) would be unsound if (3) were *Faa*. From *Faa*, that arbitrarily selected a bears F to itself, we can certainly conclude (line (5)) that something bears F to something. But we could also conclude from *Faa*

3   (4') $(\exists x)Fxx$;

yet this conclusion would not follow from (2). Given that someone is taller than an arbitrarily chosen person a, it does not follow that someone is taller than himself. *Faa* in fact is not the proper typical disjunct corresponding to (2).

120 and 121 show us in effect that the order of universal quantifiers and the order of existential quantifiers are immaterial to sense. This is *not* the case, however, with a mixture of the two quantifiers. We do have:

122 $(\exists x)(y)Fxy \vdash (y)(\exists x)Fxy$

1   (1) $(\exists x)(y)Fxy$ A

2   (2) $(y)Fay$ A

2   (3) Fab 2 UE

2   (4) $(\exists x)Fxb$ 3 EI

2   (5) $(y)(\exists x)Fxy$ 4 UI

1   (6) $(y)(\exists x)Fxy$ 1,2,5 EE

Here (2) is the typical disjunct corresponding to (1). It is essential that we select a different arbitrary name ' b ' in the application of UE at line (3). *Faa* would be a sound conclusion from (2), and we could then infer

2   (4') $(\exists x)Fxa$

(if a bears F to itself, then something bears F to a). But the step of UI at line (5) would now be unsound, since (2) contains ' a '.

The converse sequent $(y)(\exists x)Fxy \vdash (\exists x)(y)Fxy$, however, is not derivable. Nor should we wish it to be: consider the universe of people, and let F be the parent relation; then everyone has someone as parent, but it is false that there is someone who is everyone's parent. It is instructive to see how natural attempts to prove this are blocked by the restrictions on our rules. For example:

1	(1) $(y)(\exists x)Fxy$	A
1	(2) $(\exists x)Fxa$	1 UE
3	(3) Fba	A
3	(4) $(y)Fby$	3 UI?
3	(5) $(\exists x)(y)Fxy$	4 EI
1	(6) $(\exists x)(y)Fxy$	2,3,5 EE

The only faulty step is step (4)—faulty because (3) contains ' a ', so that the restriction on UI is violated. This ' near miss ' should inculcate respect for the practice of restriction observance.

There is a famous and simple argument, cited by de Morgan as an example of a kind of reasoning which, though patently sound, could not be handled within the framework of traditional logic. It runs

(1) All horses are animals; therefore all horses' heads are animals' heads.

To show the validity of (1) by our rules, we must first translate into the symbolism of the predicate calculus. Let F be being a horse, G be being an animal, and H the relation of being a head of. Then the premiss of the argument is evidently $(x)(Fx \rightarrow Gx)$. As a first step towards rendering the conclusion, we may adopt

(2) Anything that is a head of a horse is a head of an animal.

For something to be a head of a horse there must be some horse of which it is the head; in symbols, a is a head of a horse if and only if $(\exists y)(Fy \ \& \ Hay)$. Similarly, a is a head of an animal if and only if $(\exists y)(Gy \ \& \ Hay)$. The sequent, therefore, which we need to prove to demonstrate the validity of (1) is

123 $(x)(Fx \rightarrow Gx) \vdash (x)((\exists y)(Fy \ \& \ Hxy) \rightarrow (\exists y)(Gy \ \& \ Hxy))$

1	(1) $(x)(Fx \rightarrow Gx)$	A
2	(2) $(\exists y)(Fy \ \& \ Hay)$	A

3	(3) Fb & Hab	A
3	(4) Fb	3 &E
3	(5) Hab	3 &E
1	(6) $Fb \rightarrow Gb$	1 UE
1,3	(7) Gb	4,6 MPP
1,3	(8) Gb & Hab	5,7 &I
1,3	(9) $(\exists y)(Gy$ & $Hay)$	8 EI
1,2	(10) $(\exists y)(Gy$ & $Hay)$	2,3,9 EE
1	(11) $(\exists y)(Fy$ & $Hay) \rightarrow (\exists y)(Gy$ & $Hay)$	2,10 CP
1	(12) $(x)((\exists y)(Fy$ & $Hxy) \rightarrow (\exists y)(Gy$ & $Hxy))$	11 UI

Given (1) as assumption, to prove the universal proposition as conclusion we aim for the corresponding assertion concerning an arbitrarily selected object a, as at line (11). This being a conditional, we assume (line (2)) its antecedent, and aim for its consequent (line (10)). Since assumption (2) is an existential proposition, we assume the corresponding typical disjunct (line (3)), and aim for the same conclusion from that. Note that in the typical disjunct we select a new arbitrary name ' b '. The remaining strategy remains at the level of the propositional calculus until line (9), where we use EI. The step of EE is sound (line (10)), because the conclusion lacks ' b ' (though it does contain ' a '). The step of UI is sound (line (12)), because (1) lacks ' a '.

Consider next the following rather more complex argument:

> (3) Some boys like all girls; no boys like any bookworm; therefore no girls are bookworms.

The first premiss affirms that there is something which is a boy and likes all girls, that is, anything which is a girl it likes; in symbols, using F for being a boy, G for being a girl, and H for liking:

> (4) $(\exists x)(Fx$ & $(y)(Gy \rightarrow Hxy))$.

The second premiss affirms that anything which is a boy is such that it does not like any bookworm, that is, such that anything which is a bookworm it does not like; in symbols, using B for being a bookworm:

(5) $(x)(Fx \rightarrow (y)(By \rightarrow -Hxy))$.

The validating sequent which we need to prove is therefore

124 $(\exists x)(Fx \& (y)(Gy \rightarrow Hxy))$, $(x)(Fx \rightarrow (y)(By \rightarrow -Hxy))$
$\vdash (x)(Gx \rightarrow -Bx)$

1	(1) $(\exists x)(Fx \& (y)(Gy \rightarrow Hxy))$	A
2	(2) $(x)(Fx \rightarrow (y)(By \rightarrow -Hxy))$	A
3	(3) $Fa \& (y)(Gy \rightarrow Hay)$	A
3	(4) Fa	3 &E
3	(5) $(y)(Gy \rightarrow Hay)$	3 &E
2	(6) $Fa \rightarrow (y)(By \rightarrow -Hay)$	2 UE
2,3	(7) $(y)(By \rightarrow -Hay)$	4,6 MPP
8	(8) Gb	A
3	(9) $Gb \rightarrow Hab$	5 UE
3,8	(10) Hab	8,9 MPP
2,3	(11) $Bb \rightarrow -Hab$	7 UE
3,8	(12) $--Hab$	10 DN
2,3,8	(13) $-Bb$	11,12 MTT
2,3	(14) $Gb \rightarrow --Bb$	8,13 CP
2,3	(15) $(x)(Gx \rightarrow -Bx)$	14 UI
1,2	(16) $(x)(Gx \rightarrow -Bx)$	1,3,15 EE

With an existential proposition as assumption (1), we naturally assume the corresponding typical disjunct at line (3), and aim to derive the conclusion from that: hence the final step of EE. Lines (4)–(7) are concerned merely with 'itemizing' the conjunction at (3) by &E and drawing the most obvious consequence of (4), Fa, at line (7). To prove $(x)(Gx \rightarrow -Bx)$, we aim for $Gb \rightarrow -Bb$, hence the penultimate step is UI. So we assume Gb at line (8), and aim for $-Bb$. The central part of the proof is mainly propositional calculus reasoning, using the universal propositions at lines (5) and (7) in particular application to b. It is a very general tactic of discovery to proceed 'from both ends' in this manner; the central part of the desired proof will tend to be propositional in character, and relatively easy.

Consider, thirdly, the following argument:

> (6) Some botanists are eccentrics; some botanists do not like any eccentric; therefore some botanists are not liked by all botanists.

Using *F* for being a botanist, *G* for being an eccentric, and *H* for liking, we have respectively for the two premisses of (6)

(7) $(\exists x)(Fx \,\&\, Gx)$;

(8) $(\exists x)(Fx \,\&\, (y)(Gy \rightarrow -Hxy))$.

The conclusion affirms that there is something which is a botanist and which is not liked by all botanists, that is, for which it is not that the case all botanists like it. In symbols

(9) $(\exists x)(Fx \,\&\, -(y)(Fy \rightarrow Hyx))$.

We shall naturally, in seeking a proof, assume the two typical disjuncts corresponding to (7) and (8)

(10) *Fa & Ga*

(11) $Fb \,\&\, (y)(Gy \rightarrow -Hby)$.

This gives effectively four items of information, and, as a special case of the universal proposition $(y)(Gy \rightarrow -Hby)$, we have $Ga \rightarrow -Hba$, whence $-Hba$ by MPP. Now intuitively we are seeking something which is a botanist (*F*) and not liked by all botanists; *a* is such a thing, since *Fa* and *b*, who is a botanist, does not like *a*. So we aim to prove

(12) $Fa \,\&\, -(y)(Fy \rightarrow Hya)$.

The first conjunct of (12) is immediate from (10). The second can readily be proved by RAA; for, assuming $(y)(Fy \rightarrow Hya)$, we have as a special case $Fb \rightarrow Hba$, whence *Hba* contradicting $-Hba$. This intuitive discovery becomes formalized as follows:

125 $(\exists x)(Fx \,\&\, Gx), (\exists x)(Fx \,\&\, (y)(Gy \rightarrow -Hxy))$
$\vdash (\exists x)(Fx \,\&\, -(y)(Fy \rightarrow Hyx))$

1	(1) $(\exists x)(Fx \,\&\, Gx)$	A
2	(2) $(\exists x)(Fx \,\&\, (y)(Gy \rightarrow -Hxy))$	A
3	(3) *Fa & Ga*	A

4	(4) Fb & $(y)(Gy \rightarrow -Hby)$	A
3	(5) Fa	3 &E
3	(6) Ga	3 &E
4	(7) Fb	4 &E
4	(8) $(y)(Gy \rightarrow -Hby)$	4 &E
4	(9) $Ga \rightarrow -Hba$	8 UE
3,4	(10) $-Hba$	6,9 MPP
11	(11) $(y)(Fy \rightarrow Hya)$	A
11	(12) $Fb \rightarrow Hba$	11 UE
4,11	(13) Hba	7,12 MPP
3,4,11	(14) Hba & $-Hba$	10,13 &I
3,4	(15) $-(y)(Fy \rightarrow Hya)$	11,14 RAA
3,4	(16) Fa & $-(y)(Fy \rightarrow Hya)$	5,15 &I
3,4	(17) $(\exists x)(Fx$ & $-(y)(Fy \rightarrow Hyx))$	16 EI
2,3	(18) $(\exists x)(Fx$ & $-(y)(Fy \rightarrow Hyx))$	2,4,17 EE
1,2	(19) $(\exists x)(Fx$ & $-(y)(Fy \rightarrow Hyx))$	1,3,18 EE

Here, lines (3) and (4) are the typical disjuncts; lines (5)–(10) draw out the immediate consequences of (3) and (4); (11) is assumed preparatory to RAA, and the desired contradiction obtained at line (14). For EE at line (18), observe that (3), on which the conclusion rests at line (17), does not contain ' b ', though it does contain ' a '.

Finally, a rather messy, but valid, argument, which involves a predicate-letter followed by three variables:

(13) If anyone speaks to anyone, then someone introduces them; no one introduces anyone to anyone unless he knows them both; everyone speaks to William; therefore everyone is introduced to William by someone who knows him.

Let us use ' $Iabc$ ' for ' a introduces b and c ', ' Fab ' for ' a speaks to b ', ' Gab ' for ' a knows b ', and ' m ' for ' William '. The first premiss is evidently

(14) $(x)(y)(Fxy \rightarrow (\exists z)Izxy)$.

Re-thinking the second assumption, we obtain 'everyone who introduces anyone to anyone knows them both', which becomes

(15) $(x)(y)(z)(Izxy \rightarrow Gzx \& Gzy)$.

The third premiss is

(16) $(x)Fxm$,

and the conclusion

(17) $(x)(\exists y)(Iyxm \& Gym)$.

The intuitive discovery of the proof is now not difficult. Let a be an arbitrarily selected person. Then from (16) we have

(18) Fam.

From (14), as a special case, we have

(19) $Fam \rightarrow (\exists z)Izam$,

whence

(20) $(\exists z)Izam$.

Suppose now that an arbitrarily selected person b introduces a to m. We have $Ibam$, whence, from (15) by MPP,

(21) $Gba \& Gbm$.

This gives

(22) $Ibam \& Gbm$,

whence

(23) $(\exists y)(Iyam \& Gym)$.

The desired conclusion will now follow by EE and UI. Formally

126 $(x)(y)(Fxy \rightarrow (\exists z)Izxy), (x)(y)(z)(Izxy \rightarrow Gzx \& Gzy),$

$(x) Fxm \vdash (x)(\exists y)(Iyxm \& Gym)$

1	(1) $(x)(y)(Fxy \rightarrow (\exists z)Izxy)$	A
2	(2) $(x)(y)(z)(Izxy \rightarrow Gzx \& Gzy)$	A
3	(3) $(x)Fxm$	A
3	(4) Fam	3 UE
1	(5) $(y)(Fay \rightarrow (\exists z)Izay)$	1 UE
1	(6) $Fam \rightarrow (\exists z)Izam$	5 UE
1,3	(7) $(\exists z)Izam$	4,6 MPP
8	(8) $Ibam$	A

2	(9) $(y)(z)(Izay \rightarrow Gza \,\&\, Gzy)$	2 UE
2	(10) $(z)(Izam \rightarrow Gza \,\&\, Gzm)$	9 UE
2	(11) $Ibam \rightarrow Gba \,\&\, Gbm$	10 UE
2,8	(12) $Gba \,\&\, Gbm$	8,11 MPP
2,8	(13) Gbm	12 &E
2,8	(14) $Ibam \,\&\, Gbm$	8,13 &I
2,8	(15) $(\exists y)(Iyam \,\&\, Gym)$	14 EI
1,2,3	(16) $(\exists y)(Iyam \,\&\, Gym)$	7,8,15 EE
1,2,3	(17) $(x)(\exists y)(Iyxm \,\&\, Gym)$	16 UI

(16) here rests on (1), (2), and (3), because (7) rests on (1) and (3), and (15) rests on (2) as well as the typical disjunct (8). All other steps are elementary.

EXERCISES

1 Prove the validity of the following sequents:

(a) $(x)(y)(z)Fxyz \vdash (z)(y)(x)Fxyz$

(b) $(x)(\exists y)(z)Fxyz \vdash (x)(z)(\exists y)Fxyz$

(c) $(\exists x)(\exists y)(z)Fxyz \vdash (z)(\exists y)(\exists x)Fxyz$

2 Show the validity of the following arguments:

(a) If it rains, no birds are happy; if it snows, some birds are happy; therefore, if it rains, it does not snow (use 'P' for 'it rains', 'Q' for 'it snows').

(b) All camels like a gentle driver; some camels do not like Mohammed; Mohammed is a driver; therefore Mohammed is not gentle.

(c) All camels are highly strung animals; some drivers like no highly strung animals; therefore some drivers do not like any camels.

(d) Some girls like William; all boys like any girl; William is a boy; therefore there is someone who both likes and is liked by William.

(e) A whale is a mammal; some fish are whales; all fish have tails; therefore some fishes' tails are mammals' tails (use 'Tab' for 'a is a tail of b').

CHAPTER 4

The Predicate Calculus 2

1 FORMATION RULES AND RULES OF DERIVATION

So far our treatment of the predicate calculus has been relatively informal, and it is now time to do for this calculus at least part of what was done for the propositional calculus in Chapter 2. In this section, I give exact formation rules for the predicate calculus and state the four rules of derivation involving quantifiers with full precision. In the next section, I discuss substitution, theorem introduction, and sequent introduction as they apply to the predicate calculus, and state, but do not prove, consistency and completeness results.

In parallel with the discussion of Chapter 2, Section 1, we begin with some ostensive definitions of the kinds of symbol with which we have to deal. The definitions of symbols in Chapter 2 are presupposed here.

First, I define a *proper name* as one of the marks

'm', 'n', ...

Secondly, I define an *arbitrary name* as one of the marks

'a', 'b', 'c', ...

Thirdly, I define an *individual variable* as one of the marks

'x', 'y', 'z', ...

Fourthly, I define a *predicate-letter* as one of the marks

'F', 'G', 'H', ...

In each of these four definitions, as in the earlier definition of a propositional variable, we are understood theoretically to have at our disposal an indefinitely large number of distinct such marks. This is shown by the addition of '. . .' to the list. We shall in general speak of 'x', 'y', 'z', . . . as plain *variables*, where no risk of confusion with *propositional* variables is possible. Indeed, for present

138

purposes, propositional variables are to be thought of as subsumed under predicate letters, as we see in the definition (below) of an atomic sentence.

Fifthly, I define *reverse*-E to be the mark

' Ǝ '.

It proves convenient to have a word which covers both proper names and arbitrary names. Hence I define sixthly a *term* to be *either a proper name or an arbitrary name*; it is important to observe that terms do *not* include variables.

Seventhly, I define a *symbol* (*of the predicate calculus*) as *either a bracket or a logical connective or a term or an individual variable or a predicate-letter* (understood to include propositional variables) *or reverse*-E. And eighthly I define a *formula* (*of the predicate calculus*) as *any sequence of symbols*.

It now remains, as with the propositional calculus, to distinguish from the totality of formulae those which we wish to count as *well-formed* or meaningful. We therefore produce a multiple-clause definition, as before, which can be viewed as giving the *formation rules* of the predicate calculus, or as stating the basic syntax of our new language.

We begin by defining an atomic sentence. Atomic sentences play the role in the predicate calculus which propositional variables play in the propositional calculus: they are the bricks out of which complex well-formed formulae are constructed. In this and subsequent definitions we use once more the device of metalogical variables to facilitate our discussion of the language. Quite a complex array of metalogical variables will be required: thus we shall use ' P ' as a metalogical variable whose range is predicate-letters; ' t ', ' t_1 ', ' t_2 ', . . . as metalogical variables whose range is terms; ' v ', ' v_1 ', ' v_2 ', . . . as metalogical variables whose range is variables; as well as other devices.

Let $t_1, \ldots, t_n$ be any n terms (not necessarily distinct), where n is greater than or equal to 0, and P be any predicate letter. Then

$$Pt_1 \ldots t_n$$

is an *atomic sentence*. In other words, an atomic sentence is a predicate-letter followed by any (finite) number of terms. Thus

' *Fa* ', ' *Gm* ', ' *Hbn* ', ' *Ganmac* '

are atomic sentences. However,

'*Fx*', '*Hxan*'

are not atomic sentences, because '*x*' is not a term. We allow in the definition that the number of terms be 0. In this limiting case, we have in fact a propositional variable. Strictly, '*F*', '*G*', '*H*' on their own will count as atomic sentences; but where a predicate-letter has no following terms, it will be convenient to use letters '*P*', '*Q*', '*R*' once more. Thus a propositional variable may be viewed as an atomic sentence in which the predicate-letter is followed by no terms.

Now we are in a position to define a *well-formed formula* (wff) *of the predicate calculus* as follows:

(*a*) any atomic sentence is a wff;

(*b*) if A is a wff, then —A is a wff;

(*c*) if A and B are wffs, then (A → B) is a wff;

(*d*) if A and B are wffs, then (A & B) is a wff;

(*e*) if A and B are wffs, then (A v B) is a wff;

(*f*) if A and B are wffs, then (A ←→ B) is a wff;

(*g*) let A(t) be a wff containing a term t, and let v be some variable not occurring in A(t); let A(v) be a formula resulting from A(t) by replacing at least one occurrence of t by v; then (v)A(v) is a wff;

(*h*) let v be some variable and A(v) be a formula as described in (*g*); then (∃v)A(v) is a wff;

(*i*) if a formula is not a wff in virtue of clauses (*a*)–(*h*), then it is not a wff.

This definition requires some discussion. It should first be compared with the definition in Chapter 2 of a wff of the propositional calculus. Our new clause (*a*) is an extension of the original clause (*a*), replacing 'propositional variable' by 'atomic sentence'. Our basic building material for wffs now includes propositional variables, but much else besides. Clauses (*b*)–(*f*) are identical in the two definitions; as a consequence, any wff of the propositional calculus is also a wff of the predicate calculus. Thus the predicate calculus as a language is an *enlargement* of the propositional calculus. Clauses (*g*) and (*h*) introduce respectively the universal and the existential quantifier, and clause (*i*) is the usual ruling-out or extremal clause.

Clauses (*g*) and (*h*) work in a mysterious way; they can be best understood by examples. Suppose we wish to show that

(1) (x) $Fx \rightarrow Gx)$

is a wff. Observe first that neither '*Fx*' nor '*Gx*' is a wff, since '*x*' is not a term but a variable. Hence '$(Fx \rightarrow Gx)$' is not a wff either; we require this to be so, because we wish all our wffs to express *propositions*, and yet '*Fx*' does not do this—a formula containing a variable will only express a proposition true or false if, loosely speaking, that variable is tied to a quantifier and not left hanging in the air. Variables are to be construed as devices for cross-reference, not names, so to say '*x* has property *F*' is not to say anything true or false because nothing is here named. However, to say '$(x)Fx$' is to say that everything has property *F*, and so is to say something true or false.

Although '$(Fx \rightarrow Gx)$' is not a wff, '$(Fa \rightarrow Ga)$' is a wff, since '*Fa*' and '*Ga*', unlike '*Fx*' and '*Gx*', are atomic sentences. In clause (*g*), let t be the term '*a*', and let A(t) be the wff '$(Fa \rightarrow Ga)$' containing this term. Let v be the variable '*x*'. Then '$(Fx \rightarrow Gx)$' is a formula which results from A(t), namely '$(Fa \rightarrow Ga)$', by replacing at least one occurrence of '*a*' by '*x*' (in fact both occurrences of '*a*' are replaced). Hence '$(Fx \rightarrow Gx)$', though not a *well-formed* formula, is an appropriate A(v) for clause (*g*). Hence by clause (*g*) the result of prefixing '(x)' to '$(Fx \rightarrow Gx)$' is a wff: in other words (1) is a wff.

In a similar way, by modifying '$(Fa \rightarrow Ga)$' to '$(Fx \rightarrow Ga)$' or '$(Fa \rightarrow Gx)$' (i.e. by changing just one occurrence of '*a*' to '*x*'), we show by clause (*g*) that

(2) $(x)(Fx \rightarrow Ga)$

and

(3) $(x)(Fa \rightarrow Gx)$

are wffs. But there is no way of showing that

(4) $(x)(Fa \rightarrow Ga)$

is a wff, since clause (*g*) in effect stipulates that v shall occur somewhere in A(v), whilst '*x*' does not occur in '$(Fa \rightarrow Ga)$'; we do not wish to regard (4) as well-formed, since the quantifier ' (x) ' controls no variable-occurrences.

Clause (g) takes us from a wff A(t), containing a certain term t, *via* a *non*-well-formed formula A(v) containing in place of t at some occurrences a variable v which is not tied to a quantifier, to a *new* wff (v)A(v) in which the occurrences of v *are* controlled by the prefixed quantifier. In this way we secure that such expressions as (4), ' $(x)P$ ', ' $(x)Fm$ ' are not well-formed.

A further stipulation of (g) is that v shall not occur in A(t). Without this stipulation,

(5) $(x)(x)Fxx$

would be well-formed. For ' Faa ' is an atomic sentence, whence by clause (g) as it stands ' $(x)Fxa$ ' is a wff (take A(v) to be ' Fxa '); taking ' $(x)Fxa$ ' as A(t), ' a ' as t, and ' x ' as v, for A(v) we should have ' $(x)Fxx$ ', whence (5) would be well-formed. We wish to rule (5) out as well-formed, because the first universal quantifier is doing no work, the two occurrences of ' x ' after ' F ' being controlled by the second quantifier. In fact (5) is not well-formed by (g), because ' x ' already occurs in ' $(x)Fxa$ ', and so cannot be used in an application of (g). We can of course select ' y ' as v, and so show that

(6) $(y)(x)Fxy$

is a wff.

As a further illustration of the use of (g), compare the two formulae

(7) $(x)(Fx \rightarrow (x)Gx)$

(8) $((x) Fx \rightarrow (x)Gx).$

Then (7) is not a wff, but (8) is. (8) can be seen to be well-formed by clause (b), since ' $(x)Fx$ ' and ' $(x)Gx$ ' are evidently wffs. (7) is barred from being a wff by the stipulation in clause (g) that v shall not occur in A(t). The only possible wffs A(t) that could yield (7) are such formulae as ' $(Fa \rightarrow (x)Gx)$ ', ' $(Fm \rightarrow (x)Gx)$ ', which already contain the variable ' x '. Of course, by selecting a new variable, say ' y ', we can show that

(9) $(y)(Fy \rightarrow (x)Gx)$

is a wff.

Although formulae such as 'Fx', '$(Fy \rightarrow (x)Gx)$', 'Hxy' do not count as wffs by our present definition, because they are not complete sentences expressing propositions true or false in view of the 'loose' variables 'x' and 'y', it is useful in what follows to have a label for them. We shall follow logical tradition, and use the epithet 'propositional function'.[1] To be precise, a formula A is a *propositional function in the variables* $v_1, \ldots, v_n$, for n greater than or equal to 0, if $(v_1) \ldots (v_n)$A is a wff. Thus 'Hxy' is a propositional function in 'x' and 'y' because '$(x)(y)Hxy$' is a wff; '$(Fy \rightarrow (x)Gx)$' is a propositional function in 'y' because (9) is a wff; but '$(Fx \rightarrow (x)Gx)$' is *not* a propositional function in 'x' because, as we have seen, (7) is not a wff. Briefly, formulae which result from the dropping of initial quantifiers from wffs are propositional functions. In clause (g), A(v) will be a propositional function in v. By allowing the case $n = 0$, all wffs are trivially propositional functions in no variables, so that we can use 'propositional function' as a broad label to include wffs, as we proceed to do.

Clause (h) requires no separate discussion, since it merely introduces the existential quantifier into our symbolism under exactly the same conditions as are used in the case of the universal quantifier.

We define syntactically the two quantifiers as follows: a *universal quantifier* is a left-hand bracket, followed by a variable, followed by a right-hand bracket; an *existential quantifier* is a left-hand bracket, followed by a reverse-E, followed by a variable, followed by a right-hand bracket; a *quantifier* is either a universal quantifier or an existential quantifier. Then the notion of *scope* may be carried over from the propositional calculus to the predicate-calculus. The *scope* of an occurrence of a logical connective in a propositional function is *the shortest propositional function in which it occurs*. Thus in the wff (9) the scope of (the sole occurrence of) '$\rightarrow$' is '$(Fy \rightarrow (x)Gx)$', which is a propositional function in 'y', not a wff. Similarly, the *scope* of an occurrence of a quantifier in a propositional function is *the shortest propositional function in which it occurs*. Thus in (8) the

---

[1] A word about the word 'function' may not be amiss here: '$x + y$' is (expresses) a *numerical* function of x and y, since, for given numbers x and y, $x + y$ is a particular *number*; 'Fxy' is a *propositional* function in x and y in the sense that, for given individuals x and y from some universe of discourse, Fxy is a particular *proposition*.

scope of the first ' (x) ' is ' $(x)Fx$ ', and the scope of the second ' (x) ' is ' $(x)Gx$ '; in (9) the scope of (the sole occurrence of) ' (y) ' is the whole wff (9); in (6) the scope of ' (y) ' is the whole wff (6), but the scope of ' (x) ' is ' $(x)Fxy$ '—in this case a propositional function, not a wff; in the propositional function ' $(Fy \rightarrow (x)Gx)$ ', the scope of ' (x) ' is the wff ' $(x)Gx$ '.

Using the notion of scope, we may state the effect of clauses (*g*) and (*h*) as follows:

> (i) the scope of any occurrence of a quantifier in a wff or propositional function will contain *at least two* occurrences of the variable in question (one occurrence being in the quantifier itself);
>
> (ii) the scope of any occurrence of a quantifier in a wff or propositional function will not contain any other quantifier using the *same* variable.

It is in view of (i) that (4), for example, is not well-formed; for there is only one occurrence of ' x ', in the quantifier itself. And in view of (ii) (5) and (7) are not well-formed, since in the putative scope of the first ' (x) ' in each case there is a further ' (x) '. (8) escapes this stigma, because neither of the scopes of the two quantifiers is included in the other. The motive for requiring (i) and (ii) should by now be clear: in virtue of (i), every quantifier controls *some* occurrence of its variable, i.e. it is never *vacuously* used; in virtue of (ii), *every* occurrence of a quantifier's variable within its scope is controlled by that quantifier and not some other. To secure (i) and (ii), we pay a certain price in the complexity of the formation rules; our gain is a compensating simplicity in the statement of the rules of derivation and further derived rules.

As to brackets, it should be observed that clauses (*g*) and (*h*) are like clause (*b*) for ' $-$ ' in not requiring encircling brackets. Ambiguity is eliminated by the bracket-requirements in clauses (*c*)–(*f*). Thus (9), in which the scope of ' (y) ' is the whole wff, should be contrasted with

(10) $((y)Fy \rightarrow (x)Gx)$,

where the scope of ' (y) ' is merely ' $(y)Fy$ '. (9) expresses a universal proposition, whilst (10) expresses a conditional proposition whose antecedent and consequent are universal propositions. The distinction between them is secured by the brackets required for '$\rightarrow$' in clause (*c*). In practice as opposed to theory, we allow ourselves, as before, to drop outermost brackets, and to drop inside brackets where possible by virtue of the ranking of propositional connectives introduced in Chapter 2, Section 1.

Finally, given wffs $A_1, \ldots, A_n$, B of the predicate calculus, we say that

$$A_1, \ldots, A_n \vdash B$$

is a *sequent-expression* of the predicate calculus.

With the syntax of the predicate calculus accurately formulated, we may state its rules of derivation fairly easily. First, it should be pointed out that the ten primitive rules of derivation for the propositional calculus are taken over *in toto*, and now understood to apply to wffs of the predicate calculus. This secures at once that any derivable propositional calculus sequent is also a derivable predicate calculus sequent. For the four special predicate calculus rules, we proceed as follows:

UE *and* EI: let A(v) be a propositional function in v, and t be a term; let A(t) be the result of replacing all and only occurrences of v in A(v) by t. Then, given (v)A(v), UE permits us to draw the conclusion A(t). And, given A(t), EI permits us to draw the conclusion ($\exists$v)A(v). The conclusion in each case depends on the same assumptions as the premiss.

UI *and* EE: let A(e) be a wff containing the arbitrary name e, and v be a variable not occurring in A(e); let A(v) be the propositional function in v which results from replacing all and only occurrences of e in A(e) by v. Then, given A(e), UI permits us to draw the conclusion (v)A(v), provided that e occurs in no assumption on which A(e) rests. The conclusion rests on the same assumptions as the premiss. And given ($\exists$v)A(v), together with a proof of some wff C from A(e) as assumption, EE permits us to draw the conclusion C, provided that e does not occur in C or in any assumption used to derive C from A(e) (apart from in A(e) itself). The conclusion C rests on any assumptions on which ($\exists$v)A(v) depends or which are used to derive C from A(e) (apart from A(e)).

As to UE and EI, we must first observe that if A(v) is a propositional function in v, then both (v)A(v) and A(t), as defined, will be *well-formed*. For example, taking v as ' *x* ', A(v) as ' *Fx $\rightarrow$ Gxam* '

145

and t as ' a ', (v)A(v) is ' $(x)(Fx \rightarrow Gxam)$ ' and A(t) is ' $Fa \rightarrow Gaam$ ', both of which are wffs. Notice also that it does not matter if t already occurs in A(v), as here ' a ' already occurs in ' $Fx \rightarrow Gxam$ '. It is evidently quite sound to infer ' $Fa \rightarrow Gaam$ ' from ' $(x)(Fx \rightarrow Gxam)$ '. Notice, thirdly, that t must replace v at *all* its occurrences, if only because otherwise the result will not be well-formed. And notice, finally, that the control on EI is exactly the control on UE, except *in reverse*: that is, to see whether a step of EI is sound, we may consider it in reverse order as a step of UE and check that. For example, to pass from ' $Fa \& Gb$ ' to ' $(\exists x)(Fx \& Gx)$ ' is an unsound step of EI, just because to pass from ' $(x)(Fx \& Gx)$ ' to ' $Fa \& Gb$ ' would be an unsound step of UE. (Why?) The two rules are symmetrical in this respect.

As to UI and EE, we must first observe that, since v does not occur in A(e), A(v) will be a propositional function in v, since it will contain no quantifiers with v in them but will contain v. Hence both (v)A(v) and ($\exists$v)A(v) will be wffs. In applying UI, therefore, we need to select some variable not already present in the premiss: we cannot pass from ' $Fa \rightarrow (x)Gx$ ' to ' $(x)(Fx \rightarrow (x)Gx)$ ', because the supposed conclusion is not well-formed, though we can pass, if the restrictions are observed, to ' $(y)(Fy \rightarrow (x)Gx)$ '. Conversely, when we wish to use EE, to obtain a suitable typical disjunct corresponding to ($\exists$v)A(v), we should select an arbitrary name e *not already occurring* in (v)A(v), and put e in place of v at *all and only* its occurrences in the propositional function A(v). In this way, we obtain A(e) such that v does not occur in it, and such that when v is put back in place of e we again obtain the original A(v). For example, given as ($\exists$v)A(v) the wff ' $(\exists x)(Fa \rightarrow Gxb)$ ', we select e as ' c ', *not* as ' a ' or ' b ', and assume ' $Fa \rightarrow Gcb$ ' as A(e). Then A(v) for v as ' x ' will indeed be the propositional function ' $Fa \rightarrow Gxb$ '. It is perhaps worth remarking that UI and EE, like UE and EI, exhibit a certain symmetry, in that A(e) is an appropriate typical disjunct for ($\exists$v)A(v) in just the case that (v)A(v) follows from A(e) by UI.

The reader should observe that the restrictions here given for UI and EE are just the restrictions we imposed intuitively in the last chapter. He should also satisfy himself that the applications of all four rules made in the last chapter are correct in the light of their precise formulation here; a sampling of the more difficult cases will suffice.

EXERCISES

1　For each of the following formulae, state whether it is a wff, a propositional function not a wff, or neither. In case it is a wff, give a demonstration of this from the definition of wff.

(*a*) (*x*)*Gxa*

(*b*) (*x*)*Gya*

(*c*) (*x*)*Gxy*

(*d*) (∃*x*)(*Fa* & *Gx*)

(*e*) (∃*y*)(*Fx* & *Gxy*)

(*f*) (*x*)(∃*y*)(∃*z*)(*Fy* ∨ *Gz* → *Hayz*)

(*g*) (∃*x*)(∃*y*)(*Fx* ∨ *Gy* → (∃*z*)*Hayz*)

(*h*) (∃*x*)(∃*y*)(*Fx* → (*z*)(*Gz* → (∃*x*)*Hxyz*))

(*i*) (∃*x*)(*Fx* → (*z*)(*Gz* → (∃*u*)*Hxyu*))

(*j*) (∃*y*)((∃*x*) *Fx* → (*z*)(*Gz* → (∃*x*)*Hxyu*))

2　For each of the following proposed applications of UE, state whether it is correct or incorrect, and, if incorrect, why.

(*a*) 1　(1) (*x*)(∃*z*)(*Fxz* & *Gxz*)　A
　　 1　(2) (∃*z*)(*Faa* & *Gaz*)　　 1 UE

(*b*) 1　(1) (*x*)(∃*z*)(*Fxz* & *Gxz*)　A
　　 1　(2) (∃*z*)(*Faz* & *Gbz*)　　 1 UE

(*c*) 1　(1) (*x*)(∃*z*)(*Fxz* & *Gxz*)　A
　　 1　(2) (∃*z*)(*Fbz* & *Gbz*)　　 1 UE

3　For each of the following proposed applications of EI, state whether it is correct or incorrect, and, if incorrect, why.

(*a*) 1　(1) *Fba*　　　　 A
　　 1　(2) (∃*y*)*Fby*　　 1 EI

(*b*) 1　(1) *Fba*　　　　 A
　　 1　(2) (∃*x*)*Fxx*　　 1 EI

(*c*) 1　(1) *Fba*　　　　 A
　　 1　(2) (∃*y*)*Fya*　　 1 EI

(*d*) 1　(1) *Fba*　　　　 A
　　 1　(2) (∃*x*)*Fxb*　　 1 EI

(*e*) 1　(1) (∃*x*)*Fxa*　　 A
　　 1　(2) (∃*y*)(∃*x*)*Fxy*　 1 EI

147

(*f*) 1   (1) ($\exists x$)*Fxa*   A

       (2) ($\exists y$)($\exists x$)*Fxx*   1 EI

4 For each of the following proposed applications of UI, state whether it is correct or incorrect, and, if incorrect, why; assume that neither ' *a* ' nor ' *b* ' occurs in assumption (1).

(*a*) 1   (3) *Fab* ⇸ (*x*)*Gax*

    1   (4) (*y*)(*Fyb* ⇸ (*x*)*Gyx*)   3 UI

(*b*) 1   (3) *Fab* ⇸ (*x*)*Gax*

    1   (4) (*x*)(*Fxb* ⇸ (*x*)*Fxx*)   3 UI

(*c*) 1   (3) *Fab* ⇸ (*x*)*Gax*

    1   (4) (*y*)(*Fay* ⇸ (*x*)*Gax*)   3 UI

(*d*) 1   (3) *Fab* ⇸ (*x*)*Gax*

    1   (4) (*y*)(*Fyy* ⇸ (*x*)*Gyx*)   3 UI

5 For each of the following pairs of wffs, state whether the second is an appropriate typical disjunct for the first in an application of EE, and, if inappropriate, why.

(*a*) (i) ($\exists x$)(*Fxa* & (*y*)*Gby*)

    (ii) *Fba* & (*y*)*Gby*

(*b*) (i) ($\exists x$)(*Fxa* & (*y*)*Gby*)

    (ii) *Fca* & (*y*)*Gcy*

(*c*) (i) ($\exists x$)(*Fxa* & (*y*)*Gby*)

    (ii) *Fca* & (*y*)*Gby*

(*d*) (i) ($\exists x$)(*Fxa* & *Gbx*)

    (ii) *Fca* & *Gbc*

(*e*) (i) ($\exists x$)(*Fxa* & *Gbx*)

    (ii) *Fba* & *Gbb*

(*f*) (i) ($\exists x$)(*Fxa* & *Gbx*)

    (ii) *Fbm* & *Gbm*

2 SUBSTITUTION, DERIVED RULES, CONSISTENCY, AND COMPLETENESS

The notion of a theorem of the predicate calculus is analogous to that of a theorem of the propositional calculus. A *theorem* is *the conclusion of a provable sequent of the predicate calculus in which the number of assumptions is zero.* It follows at once that all theorems of the propositional calculus are theorems in the broader sense. It should also be obvious that, to each sequent proved in the previous chapter, there is a corresponding conditional provable as a theorem by supplementing the given proof with steps of CP. For example, corresponding to the sequents 103 and 105 there are the theorems:

127 $\vdash (x)(Fx \rightarrow Gx) \rightarrow ((x)Fx \rightarrow (x)Gx)$;

128 $\vdash (x)(Fx \rightarrow Gx) \rightarrow ((\exists x)Fx \rightarrow (\exists x)Gx)$.

Other theorems, corresponding at the predicate calculus level to the laws of non-contradiction (37), identity (38), and excluded middle (44), all of whose proofs are easy, are:

129 $\vdash (x) - (Fx \, \& - Fx)$;

130 $\vdash (x)(Fx \rightarrow Fx)$;

131 $\vdash (x)(Fx \vee - Fx)$.

As in the propositional calculus, theorems here may be thought of as conveying logical truths, propositions true simply on logical grounds. An important property which propositional calculus theorems were seen earlier to have is that they remain logical truths *whatever* propositions are selected in place of $P, Q, R, \ldots$; we embodied this fact in the principle of substitution (S1), that any substitution-instance of a theorem was a theorem. It is natural here to suppose that predicate calculus theorems will remain logical truths whatever *properties* are selected in place of $F, G, H, \ldots$. So we seek first for an appropriate extension of the notion of substitution-instance which will cover *substitution for predicate-letters in general*, not merely propositional variables.

We may conveniently think of properties as expressed by *propositional functions*. Thus the property F is expressed by the propositional function ' Fx ' in ' x '; the property of being both F and G by the propositional function ' $Fx \, \& \, Gx$ ' in ' x '; the property of bearing relation F to everything by the propositional function ' $(y)Fxy$ ' in ' x '. Similarly, we may think of relations as expressed by propositional functions in more than one variable. Thus ' $(\exists z)(Gxz \, \& \, Gzy)$ ' in ' x ' and ' y ' expresses the complex relation of bearing the relation G to something which bears G to. (If ' G ' is taken to be ' parent of ', then the new relation is ' grandparent of '.) Then the problem of substitution is the problem of how to replace within wffs predicate-letters followed by a certain number of terms or variables systematically by propositional functions in the same number of variables.

A simple example should make this clear. Consider theorem 131; then if ' Fx ' at its two occurrences is replaced by *any* propositional function in ' x ', we shall want to regard the result as a substitution-instance. Thus, selecting ' $(Fx \& Gx)$ ', we obtain

$$(1) \quad (x)((Fx \& Gx) \vee -(Fx \& Gx)),$$

or, selecting ' $(y)Fxy$ ', we obtain

$$(2) \quad (x)((y)Fxy \vee -(y)Fxy)),$$

and both these should convey logical truths as much as 131 does, since we expect 131 to be true independently of the actual choice of the property F.

However, certain difficulties are involved in stating the notion of substitution-instance in full generality. Consider, first, the readily proved theorem $(x)Fx \rightarrow (y)Fy$ (compare 115). If we wish to replace F by the property of being both F and G, we shall need to put ' $(Fx \& Gx)$ ' in place of ' Fx ', but ' $(Fy \& Gy)$ ' in place of ' Fy '. Or consider the theorem $(x)Fx \rightarrow Fa$. With the same replacement in mind, we here need to put ' $(Fx \& Gx)$ ' for ' Fx ', but ' $(Fa \& Ga)$ ' for ' Fa '. In other words, in substitution for a predicate-letter we shall not simply be putting *the same* formula at each of its occurrences; what we put will in part depend on what terms or variables *follow* the predicate-letter. Second, in order to obtain *well-formed* formulae after substitution, we may need to change certain variables occurring in quantifiers in the given propositional function. For example, replacing ' F ' in ' $(x)Fx \rightarrow (y)Fy$ ' by ' $(y)Fxy$ ' would yield ' $(x)(y)Fxy \rightarrow (y)(y)Fyy$ ', which is not well-formed in view of the reduplicated ' (y) '. Hence we should use the propositional function ' $(z)Fxz$ ', which clearly expresses the same property, to avoid the variable-clash. This gives the correct ' $(x)(z)Fxz \rightarrow (y)(z)Fyz$ '.

Bearing these facts in mind, let us now define quite generally a substitution-instance. It will help to have some temporary label for both terms and variables: let us call them simply *letters*; then ' a ', ' b ', ' m ', ' n ', ' x ', ' y ', etc., are all letters.

Let A be a wff of the predicate calculus, containing (occurrences of) the predicate-letter P followed by n letters. Let $Q(v_1, \ldots, v_n)$ be a propositional function in n distinct variables, $v_1, \ldots, v_n$, such that no variable in a quantifier in $Q(v_1, \ldots, v_n)$ occurs in A. For any set of n letters, $1_1, \ldots, 1_n$, let $Q(1_1, \ldots, 1_n)$ be the result of replacing v_1 by 1_1, v_2 by 1_2, $\ldots$, v_n by 1_n, throughout $Q(v_1, \ldots, v_n)$. Let A′ result from A by replacing each occurrence in A of Pl$_1 \ldots 1_n$, for letters $1_1, \ldots, 1_n$, by $Q(1_1, \ldots 1_n)$. Then A′ is said to *result from* A *by substitution*.

A′ results from A by substitution, roughly, in case *one* predicate-letter in A is appropriately replaced throughout by expressions obtained from a certain propositional function. In order to allow for *multiple* substitution on predicate-letters in A, we say that A′ is a *substitution-instance* of A in case there is a sequence of wffs such that A is the first, each results from its predecessor in the sequence by substitution, and A′ is the last. Thus A′ is a substitution-instance of A in case it results by substitution from some wff which in turn results by substitution from some wff . . . which in turn results by substitution from A. As a limiting case, we may allow trivially A to count as a substitution-instance of itself.

I illustrate this definition by one example which is as complex as any likely to be met with in practice. Let A be

(3) $(x)(Fxa \to Fba) \to (\exists y)((x)Fxx \to Fby)$

and consider the propositional function

(4) $(u)(Fuyz \ \& \ Gzya)$

in variables ' y ' and ' z '. Let P be the predicate-letter ' F ' in (3), followed by two letters. Then (4) is an appropriate $Q(v_1, \ldots, v_n)$ for the substitution, since it is a propositional function in two variables ' y ' and ' z ', and its only variable in a quantifier, namely ' u ', does not occur in (3). There are four occurrences of ' F ' in (3) to consider, namely:

(i) Fxa

(ii) Fba

(iii) Fxx

(iv) Fby.

The four corresponding versions of the propositional function (4) are obtained by replacing ' y ' and ' z ' throughout (4) by, respectively, the first and the second letter occurring after ' F '. This yields (i)′ (u)(*Fuxa* & *Gaxa*)

(ii)′ (u)(*Fuba* & *Gaba*)

(iii)′ (u)(*Fuxx* & *Gxxa*)

(iv)′ (u)(*Fuby* & *Gyba*).

Thus (i)′–(iv)′ are Q(1_1, . . ., 1_n) for the Pl$_1$ . . . 1_n given by (i)–(iv). If we now replace in (3) (i)–(iv) by (i)′–(iv)′, we obtain

(5) (x)((u)(*Fuxa* & *Gaxa*) $\rightarrow$ (u)(*Fuba* & *Gaba*)) $\rightarrow$
      ($\exists y$)((x)(u)(*Fuxx* & *Gxxa*) $\rightarrow$ (u)(*Fuby* & *Gyba*)).

Then (5), by the given definition, results from (3) by substitution, and so is a substitution-instance of (3).

For the case where $n = 0$, the predicate-letter P is simply a propositional variable, and the propositional function Q(v_1, . . ., vn) simply a wff. The replacement then consists in systematically putting some wff at all occurrences of some propositional variable in A, and the notion of substitution-instance collapses into that already defined for the propositional calculus.[1] Hence our earlier notion is no more than a special case of our present one.

We are now in a position to state a principle of substitution for the predicate calculus:

(S′1) A proof can be found for any substitution-instance of a proved theorem.

This principle is far from evident, and not easy to prove. It is beyond the scope of this book to prove it here. (The interested reader should consult, e.g., Church [2], § 35.) But a proof would consist in showing, first, that a substitution-instance is always a *wff* (not a trivial result), and, second, that a proof of A can be modified into a proof of any substitution-instance of A in such a way that applications of the four quantifier rules remain correct applications. This is not in general altogether easy to show, because of the restrictions on the rules UI and EE. The new proof may well involve a change in the arbitrary names employed, because an arbitrary name absent from the original proof may be introduced in the substitution-instance.

---

[1] However, we still need, at the predicate calculus level, to observe the restriction that the wff substituted shall contain no individual variable already occurring in A: without this restriction the result might not be well-formed—for example, replacing ' *P* ' in ' (x)(*Fx* $\rightarrow$ *P*) ' by ' (x)*Gx* ' leads to an ill-formed formula.

Finally, the notion of substitution-instance must be extended to sequents in general. This is done most simply *via* the notion of a corresponding conditional. As before, if

$$A_1, \ldots, A_n \vdash B$$

is a sequent-expression, its *corresponding conditional* is the wff

$$A_1 \to (\ldots (A_n \to B) \ldots).$$

Then we say that

$$A_1, \ldots, A_n' \vdash B'$$

is a *substitution-instance* of

$$A_1, \ldots, A_n'' \vdash B$$

if the corresponding conditional of the former sequent-expression is a substitution-instance in the sense defined of the corresponding conditional of the latter sequent-expression. This leads to the following broad principle of substitution, which includes (S'1) as a special case:

> (S'2) A proof can be found for any substitution-instance of a proved sequent.

With this labour behind us, the derived rules TI and SI are immediately forthcoming for the predicate calculus.[1] They can simply be transcribed from Chapter 2, and their justification at the new level presents no new difficulties. As in the propositional calculus, these derived rules greatly shorten the burden of proof.

These derived rules are illustrated in the following proofs:

132 $(x)Fx \vdash (x)(Gx \to Fx)$

1	(1) $(x)Fx$	A
1	(2) Fa	1 UE
1	(3) $Ga \to Fa$	2 SI(S) 50
1	(4) $(x)(Gx \to Fx)$	3 UI

133 $(x)-Fx \vdash (x)(Fx \to Gx)$

1	(1) $(x)-Fx$	A
1	(2) $-Fa$	1 UE
1	(3) $Fa \to Ga$	2 SI(S) 51
1	(4) $(x)(Fx \to Gx)$	3 UI

[1] We in fact allowed ourselves the use of them on occasion in Chapter 3, but only in connection with quite elementary propositionalus calcul sequents.

134 $(\exists x)Fx \rightarrowtail (\exists x)Gx \vdash (\exists x)(Fx \rightarrowtail Gx)$

1	(1) $(\exists x)Fx \rightarrowtail (\exists x)Gx$	A
2	(2) $-(\exists x)(Fx \rightarrowtail Gx)$	A
2	(3) $(x)-(Fx \rightarrowtail Gx)$	2 SI(S) 3.4.1(d)
2	(4) $-(Fa \rightarrowtail Ga)$	3 UE
2	(5) $Fa \ \& \ -Ga$	4 SI(S) 2.2.5(g)
2	(6) Fa	5 &E
2	(7) $-Ga$	5 &E
2	(8) $(\exists x)Fx$	6 EI
2	(9) $(x)-Gx$	7 UI
1,2	(10) $(\exists x)Gx$	1,8 MPP
1,2	(11) $-(x)-Gx$	10 SI(S) 113
1,2	(12) $(x)-Gx \ \& \ -(x)-Gx$	9,11 &I
1	(13) $--(\exists x)(Fx \rightarrowtail Gx)$	2,12 RAA
1	(14) $(\exists x)(Fx \rightarrowtail Gx)$	13 DN

132 and 133 are sequents analogous to 50 and 51, the paradoxes of material implication. They might be called the *paradoxes of formal implication* (the term ' formal implication ' was coined by Russell to describe the universal quantification over a material implication: i.e. a proposition of the form ' $(x)(Fx \rightarrowtail Gx)$ '). 132 avows that, given that *everything* has F, it follows that everything with G has F (no matter what property G may be). 133 avows that, given that *nothing* has F, it follows that everything with F has G (no matter what property G may be); thus, given that there are no unicorns, it follows that all unicorns are deliriously happy, and also, for that matter, that all unicorns are desperately sad. This paradox, is, of course, only a reflection of the fact that we are using material implication ' $\rightarrowtail$ ' in our analysis of universal propositions: if Fx is false for every x, then, by the matrix for ' $\rightarrowtail$ ', $Fx \rightarrowtail Gx$ is true for every x, whatever the truth-value of Gx may be.

The proof of 134 deserves some scrutiny. Basically a proof by RAA, lines (3)–(9) are devoted to drawing out the consequences of (2). The earlier result 3.4.1(d) cited at line (3) is

$$-(\exists x)Fx \vdash (x)-Fx.$$

A simple substitution-instance of this, taking ' Fx ' as ' $Fx \rightarrow Gx$ ', is

$$-(\exists x)(Fx \rightarrow Gx) \vdash (x) - (Fx \rightarrow Gx),$$

and this is used in connection with SI to pass from (2) to (3). (3), by elementary reasoning, has the consequences (8) and (9), the first of which is the antecedent of (1) and the second of which is shown to contradict the consequent of (1): for we can pass from (10) to (11) using the result 113 with the simple substitution of ' G ' for ' F '.

So far, we have only discussed substitution for predicate-letters; but we may also on occasion require principles allowing us to exchange one variable for another, or one arbitrary name for another, or one proper name for another. Such principles are easy to state, and not difficult to prove. Let A be a wff containing a variable v, and let w be some variable *not occurring in* A. Let A be the result of replacing all and only occurrences of v in A by w. Then if A is a theorem, so is A'. (It should be obvious from the formation rules why we require that w shall not appear in A.) This exchange can be extended to sequents in general. Similarly, let A be a wff containing the term t, and let s be some term not occurring in A. Let A' be the result of replacing all and only occurrences of t in A by s. Then if A is a theorem, so is A'. Here it makes no difference whether t and s are both proper names, both arbitrary names, or one a proper name and one an arbitrary name; the reason should be clear—if A is a theorem, it should hold whatever interpretation we give to t, so that the intuitive difference between proper and arbitrary names disappears. Again, the result extends to sequents. Both these principles may be used tacitly in connection with TI and SI. For example, we may take the proof of 100 to be also a proof of

$$Fa, (y)(Fy \rightarrow Gy) \vdash Ga,$$

where ' m ' has been replaced by ' a ', and ' x ' by ' y '.

It remains in this section to sketch the notions of consistency and completeness for the predicate calculus, and so to do for the predicate calculus what was done for the propositional calculus in Chapter 2. In the propositional calculus, we described a property of sequent-expressions—that of being *tautologous*—and showed that all and only derivable sequents had this property, thus demonstrating consistency and completeness. In the case of the predicate calculus, no property which can be described in the simple language of the truth-table test is forthcoming; but an analogous, though more complex, property of wffs can be specified—that of being *true under all interpretations in every non-empty universe*.

We have appealed to the notion of a universe of discourse already. A universe is quite simply some set of objects: it may be finite, as the universe of three objects considered in Chapter 3 was; it may be infinite, as the universe of the natural numbers, considered in algebra, is. A *non-empty* universe is a universe which contains at least one object. The predicate calculus makes the assumption that, on any interpretation, we are discussing a non-empty universe; for example, we have as a *theorem* (from 131 by 104) that $(\exists x)(Fx \vee -Fx)$, though this would not be true in an empty universe, since it is an existential proposition. Indeed, empty universes have such peculiar formal properties that it is better *not* to consider them.

An *interpretation* of a wff in a given non-empty universe is an assignment of objects from the universe to the *terms* in the wff, together with an assignment of *properties* and *relations* defined for objects in the universe to the *predicate-letters* in the wff (if the predicate-letters are propositional variables, we simply assign truth-values to them, as in the propositional calculus). For example, we may interpret

(6) *Fa* & $(\exists x)Gax$

in the universe of natural numbers, by assigning 2 to ' *a* ', the property of being even to ' *F* ', and the relation of being greater than to ' *G* '. Under this interpretation, (6) affirms that 2 is even and there is a natural number than which 2 is greater; hence (6) is evidently true for this interpretation. In general we can compute the truth-value of a wff for a given interpretation in a given universe by obvious means; we take the variables to range over the objects in the universe, and ' $(x)(\ldots x \ldots)$ ' will be true in just the case that all objects in the universe satisfy the condition ' $(\ldots x \ldots)$ ', ' $(\exists x)(\ldots x \ldots)$ ' true in just the case that at least one object in the universe satisfies the condition ' $(\ldots x \ldots)$ '. For the rest, we use the propositional calculus matrices to determine the truth-values of complex sentences, given the truth-values of their components. This account is admittedly sketchy, but enough should have been said to give substance to the idea of a wff being *true under a certain interpretation in a certain non-empty universe.*

Let us now say that a wff is *logically true* if it is true under *all* interpretations in *all* non-empty universes. This is the desired extension, for the predicate calculus, of the property of being tautologous for wffs of the propositional calculus. If a given universe is finite, we can actually *list* all possible interpretations of a wff; for the number of distinct assignments of objects, properties, and relations is in this case finite. Hence for finite universes, a mechanical test, analogous to the truth-table procedure, is available for testing truth or falsity under interpretation. But clearly no such technique is available for an infinite universe. In this resides the fundamental distinction between the propositional and predicate calculi.

For sequent-expressions in general, we need the following. A sequent-expression

$$A_1, \ldots, A_n \vdash B$$

is *logically valid* if, under any interpretation in any non-empty universe of the wffs $A_1, \ldots, A_n$, B, whenever $A_1, \ldots, A_n$ are all true then so is B. It follows that a sequent-expression is logically valid if and only if its corresponding conditional is logically true.

It is not too difficult a matter to show that all derivable sequents are logically valid, and that therefore the predicate calculus is consistent. As before, this involves a case-by-case consideration of the fourteen rules, to show that no application of any of them leads from a logically valid sequent to a sequent not logically valid. To show the converse is a good deal harder. The first completeness proof for the predicate calculus was obtained by Kurt Godel in 1930. There is an account of this proof in Hilbert and Ackermann [7]; the interested reader should further consult Church [2], §§ 44 and 45. The method of proof can readily be extended to the present formulation of the predicate calculus, and we obtain the result that any logically valid sequent is derivable from our rules. This proof may be regarded as the borderline between elementary and advanced logic.

Although most of this section has imitated the procedure of Chapter 2, and most of the results concerning the predicate calculus have been extensions of results concerning the propositional calculus,

157

in the case of completeness there is an important difference. The completeness proof in Chapter 2, Section 5, is such that, given a tautologous sequent, we can actually *construct* from the proof a derivation of the sequent in question. But Gödel's completeness proof does not have this constructive character. Indeed, as was pointed out in Chapter 2, there is no mechanical way, like the truth-table test, of sorting sequent-expressions of the predicate calculus into the logically valid and the logically invalid. This result is due to Alonzo Church, and for an elementary discussion of it the reader should consult Quine [17], § 32. What this means in practice is that, faced with a certain sequent-expression and wishing to know whether it is sound or unsound, we may search imaginatively for a proof and we may search imaginatively for an interpretation in some non-empty universe which will render all assumptions true and conclusion false. A proof, if one is found, can be mechanically checked; an interpretation showing unsoundness, if one is found, can also be confirmed to be such. But the search itself for inter-pretation or proof cannot be reduced to rule. In this respect, the predicate calculus differs essentially from the propositional calculus.

EXERCISES

1 For the sequents mentioned below, carry out the given substitutions.

(*a*) In 101, for ' F ' substitute ' $Fx \vee Hx$ ', for ' G ' substitute ' Hxa '.

(*b*) In 103, for ' G ' substitute ' $(y)Gxy$ '.

(*c*) In 104, for ' F ' substitute ' $(\exists y)(Fxy \vee Gya)$ '.

(*d*) In 109, for ' F ' substitute ' $Fx \rightarrow Hx$ ', for ' G ' substitute ' $Hx \rightarrow Fx$ '.

(*e*) In 116, for ' F ' substitute ' $(\exists z)(Fxz \& Gzx)$ '.

(*f*) In 119, for ' F ' substitute ' $(y)Gyx$ ', for ' P ' substitute ' $(z)Gza$ '.

(*g*) In 120, for ' F ' substitute ' $(z)Fxyz$ ' in ' x ' and ' y '.

(*h*) In 123, for ' F ' substitute ' $Fxa \vee (z)Kzx$ ', for ' G ' substitute ' $(z)Fzx$ ', for ' H ' substitute ' $(z)(Fxz \rightarrow Gzy)$ ' in ' x ' and ' y '.

2 Using TI or SI where appropriate, show the validity of the following sequents:

(*a*) $(x)Fx \& (\exists x)Gx$ $(\exists x)(Fx \& Gx)$

(*b*) $(x)Fx \vee (\exists x)Gx \vdash (\exists x)(Fx \vee Gx)$

(*c*) $(\exists x)Fx \rightarrow (x)Gx \vdash (x)(Fx \rightarrow Gx)$

(*d*) $(x)(Fx \vee Gx) \vdash (x)Fx \vee (\exists x)Gx$

    (e) $(x)Fx \twoheadrightarrow (x)Gx \vdash (\exists x)(Fx \twoheadrightarrow Gx)$

    (f) $(\exists x)(Fx \twoheadrightarrow Gx) \dashv\vdash (x)Fx \twoheadrightarrow (\exists x)Gx$

3   Prove the following theorems:

    (a) $\vdash (y)((x)Fx \twoheadrightarrow Fy)$

    (b) $\vdash (y)(Fy \twoheadrightarrow (\exists x)Fx)$

    (c) $\vdash (\exists y)(Fy \twoheadrightarrow (x)Fx)$

    (d) $\vdash (\exists y)((\exists x)Fx \twoheadrightarrow Fy)$

4   (Perhaps the hardest exercise in the book)

    (a) The interderivability result 113 suggests that we might *define* the existential quantifier in terms of the universal quantifier and negation, in much the way that in Chapter 1 we defined ' $\leftarrow\!\!\twoheadrightarrow$ ' in terms of ' $\twoheadrightarrow$ ' and ' & '. Thus, for any variable v:

$$Df.\ \exists: \quad (\exists v) = -(v)-.[1]$$

By this definition, we understand ' $(\exists x)$ ', for example, simply as an abbreviation for ' $-(x)-$ '. Suppose we adopt this definition, and accordingly drop from the predicate calculus the rules EI and EE, leaving ourselves with only UI and UE as primitive rules (as well, of course, as the propositional calculus rules). Show that from this basis the rules EI and EE can be obtained as *derived rules*. (Hint: in the (simpler) case of EI, it suffices to show that $A(t) \vdash -(v)-A(v)$ can always be proved using only UI and UE, where $A(t)$ and $A(v)$ are as described at the end of Section 1.)

    (b) The interderivability result 114 similarly suggests a possible definition of the universal quantifier in terms of the existential, thus:

$$Df.\ U: \quad (v) = -(\exists v)-.$$

Show conversely that, if we adopt this definition together with the rules EI and EE and propositional calculus rules, the rules UI and UE can be obtained as *derived rules*.

3 IDENTITY

The remainder of this chapter is devoted to particular applications of the predicate calculus. In the present section we study one particular relation of special importance for logic, the relation of identity. This relation is already familiar from mathematics, where

[1] This procedure is in fact adopted in many accounts of the predicate calculus.

it is marked by the (misleadingly called) equality-sign ' = '. Thus the force of ' $2 + 2 = 4$ ' is that the number which results from adding two and two is (the same or identical number as) four. In non-mathematical contexts, identity is expressed usually by ' is '; but since the verb ' to be ' has many senses, we must indicate first in *which* sense ' is ' expresses identity.

Consider the six English sentences:

(1) Socrates is a philosopher;

(2) Paris is a city;

(3) Courage is a virtue;

(4) Socrates is the philosopher who taught Plato;

(5) Paris is the capital of France;

(6) Courage is the virtue I most admire;

Sentences (1)–(3) are simple subject-predicate sentences; a particular object (Socrates, Paris, courage) is said to have a certain property (being a philosopher, being a city, being a virtue). We accordingly call the ' is in (1)–(3) the ' is' *of predication*. This use of ' is ' must be contrasted with the ' is ' in (4)–(6), where rather the sense is ' is the same object as ' (with ' object ' used in some broad neutral sense). This ' is ' we distinguish as the ' is ' *of identity*.

Aids towards recognizing the ' is ' of identity are: (*a*) can ' is ' be replaced by ' is the same object as '?—if so, ' is ' is ' is ' of identity, if not, not; (*b*) can the phrases flanking ' is ' on both sides be reversed preserving approximately the same sense?—if so, ' is ' is ' is ' of identity, if not, not. Applying these two tests to (1)–(6) should reveal the difference between (1)–(3) on the one hand and (4)–(6) on the other. There are indeed certain types of expression which regularly flank the ' is ' of identity on both sides: first, proper names, such as ' Napoleon ', ' Waterloo '; secondly, what grammarians sometimes call abstract nouns (as opposed to common nouns), which are distinguished by lacking a plural, such as ' courage ', ' bread ', ' oxygen '; thirdly, singular phrases beginning with ' the ', such as ' the evening star ', ' the author of Waverley '— such phrases are often called *definite descriptions*; fourthly, what we may call demonstrative words and phrases, such as ' I ', ' he ', ' that book ', ' this cloud ', ' last night '. Here are more examples of the ' is ' of identity (or in one case ' was '), flanked by such expressions:

(7) The latest element to be discovered is uranium;

(8) That tall man is his first cousin;

(9) Last night was the first night of the fair;

(10) Cicero is Tully;

(11) Beauty is truth.

A fuller discussion of identity, from a philosophical standpoint, is beyond the scope of this book. Here we are concerned with the formal handling of the notion. To this end, we adopt from mathematics the symbol ' = ' to represent the ' is ' of identity, and expand our formation rules accordingly. Let t and s be any terms; then (t = s) is now to count as an atomic sentence. Thus ' $(a = b)$ ', ' $(c = c)$ ', ' $(m = n)$ ', ' $(a = n)$ ' are all atomic sentences in the extended sense. In a way, ' = ' is a new predicate-letter controlling two terms; but, following mathematical practice, we adopt the convention of placing it *between* the two terms rather than *before* them; and, unlike other predicate-letters, it has a *fixed* interpretation. This new language, only a slight extension of the old, is *the predicate calculus with identity*. In view of the other formation rules, ' = ' can appear in complex expressions in just the way that ' F ', ' G ', ... appear. Thus ' $(x)(x = x)$ ', ' $(x)(y)((Fx \& (x = y)) \rightarrow Fy)$ ' are wffs of the extended language.

To handle identity in proofs, we introduce two simple rules, a rule of identity introduction (=I) and a rule of identity elimination (=E). For any term t, the rule =I permits us to introduce into a proof at any stage t = t, resting on no assumptions. The idea should be clear: anything is itself, as a matter of logic; hence t = t is logically true, and so can appear without assumptions. Now let t and s be terms, and A(t) a wff containing (occurrences of) t; let A(s) be the result of replacing at least one occurrence (but not necessarily all) of t in A(t) by s; then, given the premisses t = s and A(t), the rule =E permits us to draw A(s) as conclusion, resting on the pool of the assumptions on which the premisses rest. Again the idea is straightforward: if t *is* s, then, given A(t)—a proposition about t—we can infer A(s)—the corresponding proposition about s. For example, if beauty *is* truth, and beauty is in the eye of the beholder, then truth is in the eye of the beholder (an argument designed to show that Keats was wrong).

Putting these rules to work, we have:

135 $a = b \vdash b = a$

1	(1) $a = b$	A	
	(2) $a = a$	$=$I	
1	(3) $b = a$	1,2 $=$E	

136 $a = b \;\&\; b = c \vdash a = c$

1	(1) $a = b \;\&\; b = c$	A	
1	(2) $a = b$	1 &E	
1	(3) $b = c$	1 &E	
1	(4) $a = c$	2,3 $=$E	

137 $Fa \dashv\vdash (\exists x)(x = a \;\&\; Fx)$

    (a) $Fa \vdash (\exists x)(x = a \;\&\; Fx)$

1	(1) Fa	A	
	(2) $a = a$	$=$I	
1	(3) $a = a \;\&\; Fa$	1,2 &I	
1	(4) $(\exists x)(x = a \;\&\; Fx)$	3 EI	

    (b) $(\exists x)(x = a \;\&\; Fx) \vdash Fa$

1	(1) $(\exists x)(x = a \;\&\; Fx)$	A	
2	(2) $b = a \;\&\; Fb$	A	
2	(3) $b = a$	2 &E	
2	(4) Fb	2 &E	
2	(5) Fa	3,4 $=$E	
1	(6) Fa	1,2,5 EE	

The proof of 135 may be perplexing: but think of (1) ' $a = b$ ' as the premiss t = s, and (2) ' $a = a$ ' as A(t); then (3) ' $b = a$ ' is a suitable A(s), since it results from (2) by replacing the first occurrence of ' a ' in (2) by ' b ', in accordance with the identity (1). The step of $=$E in 136 is similar. According to 137, the proposition that an arbitrarily selected object a has F is interderivable with the proposition that there is something which *is a* and has property F.

As theorems concerning identity, we obtain

138 $\vdash (x)(x = x)$

    (1) $a = a$ $=$I

    (2) $(x)(x = x)$ I UI

139 $\vdash (x)(y)(x = y \rightarrow y = x)$

140 $\vdash (x)(y)(z)(x = y \mathbin{\&} y = z \rightarrow x = z)$

These last can be obtained by supplementing the proofs of 135 and 136 with steps of CP followed by steps of UI.

It is worth remarking that if, for a step of $=$E, the terms are given in the wrong order (s $=$ t rather than t $=$ s), we can obtain t $=$ s by SI using 135 (after a change of lettering if necessary).

Consider now the argument (adapted from Quine [17])

> (12) Only Smith and the guard at the gate knew the password; someone who knew the password stole the gun; therefore either Smith or the guard at the gate stole the gun.

This is evidently sound, but its soundness cannot be shown in the predicate calculus without identity. Bearing in mind the usual force of ' only ', the first premiss of (12) means

> (13) Everyone who knew the password either was Smith or was the guard at the gate.

In (13), the two ' was's ' are ' was's ' of identity. Hence, using ' K ' for knowing the password ' m ' for Smith and ' n ' for the guard at the gate, we transform (13) into

> (14) $(x)(Kx \rightarrow x = m \lor x = n)$.

Using ' S ' for stealing the gun, we must prove the sequent

141 $(x)(Kx \rightarrow x = m \lor x = n), (\exists x)(Kx \mathbin{\&} Sx) \vdash Sm \lor Sn$

1	(1) $(x)(Kx \rightarrow x = m \lor x = n)$	A
2	(2) $(\exists x)(Kx \mathbin{\&} Sx)$	A
3	(3) $Ka \mathbin{\&} Sa$	A
3	(4) Ka	3 &E
3	(5) Sa	3 &E
1	(6) $Ka \rightarrow a = m \lor a = n$	1 UE
1,3	(7) $a = m \lor a = n$	4,6 MPP

8	(8) $a = m$	A
3,8	(9) Sm	5,8 $=$E
3,8	(10) Sm v Sn	9 vI
11	(11) $a = n$	A
3,11	(12) Sn	5,11 $=$E
3,11	(13) Sm v Sn	12 vI
1,3	(14) Sm v Sn	7,8,10,11,13 vE
1,2	(15) Sm v Sn	2,3,14 EE

After assuming the typical disjunct at line (3) corresponding to line (2), we obtain (7) that either a is m or a is n. On either supposition, it follows that m stole the gun or n stole the gun. Hence by vE and EE (lines (14) and (15)) we obtain the desired conclusion.

This argument illustrates the increased expressive power we obtain by adding identity to our list of logical notions. For another example, consider first the following derivable sequent:

142 $(\exists x)Fx \vdash (\exists x)(\exists y)(Fx \,\&\, Fy)$

1	(1) $(\exists x)Fx$	A
2	(2) Fa	A
2	(3) $Fa \,\&\, Fa$	2,2 &I
2	(4) $(\exists y)(Fa \,\&\, Fy)$	3 EI
2	(5) $(\exists x)(\exists y)(Fx \,\&\, Fy)$	4 EI
1	(6) $(\exists x)(\exists y)(Fx \,\&\, Fy)$	1,2,5 EE

(The result can in fact be strengthened to an interderivability result.) It follows from the validity of this sequent that if only *one* object has F, then $(\exists x)(\exists y)(Fx \,\&\, Fy)$; in other words, when we use distinct *variables* ' x ' and ' y ' it does *not* follow that there are corresponding distinct *objects*. In order to express that there are at least two distinct objects with property F, we need the identity-symbol:

$$(\exists x)(\exists y)((Fx \,\&\, Fy) \,\&\, -(x = y))$$

—there is an x and a y both with F which are not identical.[1]

---

[1] This point is important if our use of variables is to be properly understood. If someone killed himself, then someone killed someone; this becomes for us the theorem $(\exists x)Kxx \rightarrow (\exists x)(\exists y)Kxy$. If we wish to say that someone killed someone *other than himself*, we need to write ' $(\exists x)(\exists y)(Kxy \,\&\, -(x = y))$ ', and this will *not* follow from $(\exists x)Kxx$.

Similarly, to say that there are at least three things with F we can write

$$(\exists x)(\exists y)(\exists z)(Fx \;\&\; Fy \;\&\; Fz \;\&\; -(x = y) \;\&\; -(x = z) \;\&\; -(y = z)).$$

(For the sake of clarity, I here carelessly ignore the inner bracketing of the complex conjunction.) In general, for any number n, it should be obvious how we can say that there are at least n things with F, using essentially the identity-symbol in the analysis.

If we can translate 'there are *at least n* things with F', can we perhaps translate also 'there are exactly n things with F'? Let us begin with 'there is exactly *one* thing with F' or 'there is one and *only one* thing with F' or 'there is at least one and *at most* one thing with F'. To say 'there is at least one thing with F' is simply to say $(\exists x)Fx$, so that the problem reduces to translating 'there is at most one thing with F'.

To claim that *at most* one thing has F is to claim (in idiomatic but misleading English) that any two things with F are the same thing. In our symbolism it is

(15) $(x)(y)(Fx \;\&\; Fy \rightarrow x = y)$

—take objects x and y; then if both have F they are identical. This formula allows there to be nothing with F, and one thing with F; but if there are more than one, it becomes evidently false. Hence (15) represents the claim that *at most* one thing has F.

To say, therefore, that exactly one thing has F is to say

(16) $(\exists x)Fx \;\&\; (x)(y)(Fx \;\&\; Fy \rightarrow x = y).$

Now (16) is actually interderivable with the briefer and neater

(17) $(\exists x)(Fx \;\&\; (y)(Fy \rightarrow x = y)).$

(17) affirms that something has F and anything with F *is that very thing*: another way of saying that exactly one thing has F. Yet a third equivalent, which may be clearer still, is

(18) $(\exists x)(Fx \;\&\; -(\exists y)(Fy \;\&\; -(x = y)))$

—there is something with F and *nothing else* with F.

Let us agree to use the notation '$(\exists_1 x)Fx$' as an abbreviation for, say, (17) ((16) or (18) would do as well): we may call this new symbol a *numerically definite quantifier*, and read ' there is exactly one x with F ' or ' there is a unique x with F '.

If now we wish to say that there are *exactly two* things with F, several courses are open. Perhaps the simplest is to write

(19) $(\exists x)(\exists y)(Fx \mathbin{\&} Fy \mathbin{\&} -(x = y) \mathbin{\&} (z)(Fz \rightarrow z = x \vee z = y))$

—there are distinct objects x and y, both with F, such that anything with F is either the one, x, or the other, y. Alternatively, we may use the already introduced symbol ' $(\exists_1 x)$ ', and write

(20) $(\exists x)(Fx \mathbin{\&} (\exists_1 y)(Fy \mathbin{\&} -(x = y)))$

—there is something with F such that there is exactly one thing with F not the same as it: clearly equivalent to the claim that exactly two things have F. We might wish to abbreviate (20) by using a second numerically definite quantifier ' $(\exists_2 x)Fx$ '.

This procedure can be extended to any finite number; that is, for any number n, we can find an expression in the predicate calculus with identity which asserts that exactly n things have F. We shall not pursue this development here, however, but revert finally instead to a discussion of the handling in argument of definite descriptions (singular phrases beginning with ' the ').

In argument (12) of this section appeared the definite description ' the guard at the gate '. We handled it in our translation exactly as if it had been a proper name like ' Smith ', and used ' n '. Here, we were able to show validity by this approach; but that is not always so. Consider, for example, the argument (due in essentials again to Quine [17])

(21) The author of the Iliad wrote the Odyssey; therefore someone wrote both the Iliad and the Odyssey.

If we treat ' the author of the Iliad ' as a proper name, and represent it by ' m ', say, the soundness of the argument does not emerge. The premiss becomes ' Om ' and the conclusion ' $(\exists x)(Ix \mathbin{\&} Ox)$ ', and the corresponding sequent is not derivable. Clearly, the soundness of the argument hinges here on the *property involved in the definite description*, the property of writing the Iliad, and this property will have to emerge in our analysis of the premiss if we are to show validity. Reflection on the content of the definite description in (21) suggests that the premiss can be taken to affirm that *exactly one* person wrote the Iliad and *that person* wrote the Odyssey. Hence the premiss becomes

(22) $(\exists x)((Ix \ \& \ Ox) \ \& \ (y)(Iy \rightarrow x = y))$

—someone wrote the Iliad, and wrote the Odyssey, and further that person is *unique* in having written the Iliad; the last clause expresses his uniqueness in the manner of (17) above, and so catches the force of the definite description. But from (22) the conclusion $(\exists x)(Ix \ \& \ Ox)$ follows at once by SI using 111.

The treatment of definite descriptions in (22) is of considerable importance in logical analysis; due to Russell, it has come to be known as Russell's theory of definite descriptions. It raises philosophical problems, which are, however, beyond the scope of this book. Here it should be observed that, as the contrast between (12) and (21) shows, there is nothing *obligatory* about Russell's analysis: often, in testing for validity, it suffices to consider definite descriptions entirely on a par with ordinary proper names. Roughly, we may say that how we handle definite descriptions depends on how much of the internal structure of propositions we need to reveal in order to validate arguments in which those propositions occur. And this in turn tells us something about the *logical form* of propositions. There is nothing final or absolute about our analysis of ordinary sentences into logical notation. For the purposes of revealing the validity of arguments, the same sentence may in one reasoning context be represented simply by ' P ' (if validity hinges on propositional calculus structure alone) and in another by a complex predicate-calculus wff; and similarly a definite description inside a sentence may in one context be represented simply by ' m ' and in another require to be analysed more fully by means of Russell's theory of descriptions. We can perhaps express this point by saying that the logical form of a sentence is always relative to a given arguing situation. Or perhaps it would be better not to speak of the logical form of sentences at all, but only of the logical form of arguments in which sentences are used.

EXERCISES

1 Prove the validity of the following sequents:

    (a) $Fa \dashv\vdash (x)(x = a \rightarrow Fx)$

    (b) $\vdash (x)(y)(Fx \ \& \ x = y \rightarrow Fy)$

(c) $b = a, c = a \vdash b = c$

(d) $a = b \vdash Fa \longleftrightarrow Fb$

(e) $a = b \vdash c = a \longleftrightarrow c = b$

(f) $\vdash (\exists x)(x = a)$

2   Prove the soundness of the following arguments by translating them into the symbolism of the predicate calculus with identity and showing the validity of the corresponding sequents:

    (a) All murderers are insane; Jekyll is a murderer; Jekyll is Hyde; therefore Hyde is insane.

    (b) No murderers are sane; Jekyll is a murderer; Hyde is sane; therefore Jekyll is not Hyde.

    (c) Only Tom and Jane are dancing; Tom and Jane are both doing the twist; therefore everyone dancing is doing the twist.

    (d) There is at most one unscrupulous head of state; Mao Tse-tung is an unscrupulous head of state; Johnson is not Mao Tse-tung; therefore Johnson is not an unscrupulous head of state.

3   Establish the following interderivability result (compare (16) and (17) of the text):

    (a) $(\exists x)(Fx \mathbin{\&} (y)(Fy \rightarrow x = y)) \dashv\vdash (\exists x)Fx \mathbin{\&} (x)(y)(Fx \mathbin{\&} Fy \rightarrow x = y)$

    (Hint: SI using (c) of Exercise 1 may shorten the labour.)

4   Write down an expression from the predicate calculus with identity with the meaning:

    (a) there are at most two things with F;

    (b) there are exactly three things with F.

5   Using Russell's theory of definite descriptions, establish the soundness of the following argument:

    (a) The author of *Mein Kampf* died in 1945; Hitler wrote *Mein Kampf*; therefore Hitler died in 1945.

4 THE SYLLOGISM

The predicate calculus was undiscovered 100 years ago. It owes its development to logicians working at about the turn of the present century, in particular to Gottlob Frege and to Russell. For well over 2,000 years before that, some of the same logical material was handled by the theory of the syllogism, which we owe to Aristotle; virtually nothing was added to it in that period. There can today be no doubt that predicate calculus has replaced the syllogism as an instrument for serious logical work; predicate calculus is to syllogism what a precision tool is to a blunt knife. (None the less, whenever a new piece of equipment is introduced, there will always be found those who prefer the outdated machinery with which they are familiar; and predicate calculus is unquestionably harder to learn.) There are no reasons other than historical ones for studying the syllogism; but this theory has been of importance in the history of both logic and philosophy, and perhaps therefore deserves a place in a modern logic course. Our treatment will be brisk, since the material may be found in other works (for example, Joseph [8] or Stebbing [22]); but I shall try also to clarify the relation between syllogisms and predicate calculus, about which there is dispute (see Strawson [23]).

The theory of the syllogism studies just four types of proposition, which we distinguished in Chapter 3, Section 1: (i) Everything with F has G, $(x)(Fx \rightarrow Gx)$, which we call a *universal affirmative* and symbolize $A(F, G)$; (ii) nothing with F has G, $(x)(Fx \rightarrow -Gx)$, which we call a *universal negative* and symbolize $E(F, G)$; (iii) something with F has G, $(\exists x)(Fx \& Gx)$, which we call a *particular affirmative* and symbolize $I(F, G)$; (iv) something with F has not G, $(\exists x)(Fx \& -Gx)$, which we call a *particular negative*, and symbolize $O(F, G)$. In syllogism, as opposed to predicate calculus, we analyse the structure of these propositions no further; we merely isolate the two properties involved, and record the rest in one of the symbols ' A ', ' E ', ' I ', ' O '. It is customary, but unfortunate from our standpoint, to call the letters ' F ', ' G ', . . . *terms*; in this section, I shall follow this practice and hope that the generated ambiguity is not disastrous.

The basic logical relations between the four forms are traditionally set out in a diagram, called *the square of opposition* (see page 170). The meaning of the terminology in the square of opposition was given, at least at the propositional calculus level, in Chapter 2, Section 3. In view of the discussion there, we may embody the

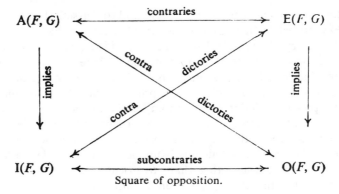

Square of opposition.

claims made in the diagram in the six principles:

(1) $-(A(F, G) \longleftrightarrow O(F, G))$ (A and O contradictories);
(2) $-(I(F, G) \longleftrightarrow E(F, G))$ (I and E contradictories);
(3) $-(A(F, G)\ \&\ E(F, G))$ (A and E contraries);
(4) $I(F, G)\ v\ O(F, G)$ (I and O subcontraries);
(5) $A(F, G) \rightarrow I(F, G)$ (A implies I);
(6) $E(F, G) \rightarrow O(F, G)$ (E implies O).

These principles are not independent of one another. For example, given (1) and (2) *by propositional calculus reasoning alone* we can deduce (4) from (3) or (3) from (4), and (5) from (6) or (6) from (5) —this is left as a pretty exercise.

Thus the traditional view is that A and O must have opposite truth-values in all circumstances, and so must I and E ((1) and (2)). A and E cannot both be true, but may both be false—indeed will be both false in case I and O are both true; conversely I and O may be both true, and will be in case A and E are both false, but cannot be both false ((3) and (4)). Whenever A is true, so is I, and whenever E is true, so is O ((5) and (6)). Consideration of simple examples will support this doctrine.

In the square of opposition, the order in which the terms ' *F* ' and ' *G* ' are given is considered to be fixed throughout. However, there are traditional principles, known as *laws of conversion*, concerned with interchange of term-order. We have in fact

(7) $I(F, G) \rightarrow I(G, F)$
(8) $E(F, G) \rightarrow E(G, F).$

The converses of (7) and (8) follow from (7) and (8) by the exchange of ' F ' and ' G ', so that both conditionals can at once be strengthened to biconditionals. (7) is known as the law of the *simple conversion* of I, and (8) as the law of the *simple conversion* of E. The conversion is simple in that an I-form converts into an I-form and an E-form into an E-form. There are no such principles for A- and O-forms; for we clearly have neither $A(F, G) \not\rightarrow A(G, F)$ nor $O(F, G) \not\rightarrow O(G, F)$. However, for A we have

$$(9)\ A(F, G) \rightarrow I(G, F).$$

((9) is a consequence of (5) and (7) by propositional calculus reasoning.) (9) is sometimes expressed by saying that the A-form *converts per accidens* into the I-form, and called the law of the *conversion per accidens* of A. The converse of (9) is not obtainable, and there is no such principle at all for the O-form. Hence I, E, and A are all said to convert, but O is said not to convert.

Other traditional laws, such as principles of *obversion*, depend on the introduction of negative terms ' $-F$ ', ' $-G$ ', etc. Such terms are not required for the Aristotelian theory of the syllogism, so we shall ignore them here. The interested reader should consult Joseph [8] or Stebbing [22].

Unlike the principles enunciated so far, a syllogism is concerned with *three* terms, say ' F ', ' G ', and ' H '. To define a syllogism precisely, let three terms P_1, P_2, and P_3 be given, and let us call P_1 the *minor term*, P_2 the *middle term*, and P_3 the *major term*. Then a syllogism is a sequence of three propositions (the first two called the *premisses* and the last the *conclusion*) each of either the A-, E-, I-, or O-form, such that the conclusion contains the minor and major terms in that order, the first premiss contains the major and middle terms, and the second premiss contains the minor and middle terms. The first premiss, since it contains the major term, is the *major* premiss, and the second premiss, since it contains the minor term, is the *minor* premiss. The middle term appears in both premisses but not the conclusion. For example,

$$(10)\ \begin{array}{l} A\ (F, H) \\ \underline{I\ (H, G)} \\ O\ (G, F) \end{array}$$

can be recognized as a syllogism, since it concerns three terms ' F ',

' *G* ', and ' *H* ' such that of the two appearing in the conclusion the first appears in just the second premiss and the second in just the first premiss (in particular, we recognize ' *F* ' as the major term since it appears in the first premiss and ' *G* ' as the minor term since it appears in the second premiss), whilst the third term appears in both premisses but not in the conclusion (hence ' *H* ' is here the middle term). We can thus take P_1 as ' *G* ', P_2 as ' *H* ' and P_3 as ' *F* ', and the definition is satisfied. This example incidentally determines a convention for writing syllogisms.

Consider now

(11) $A(H, F)$
$\underline{I(F, G)}$
$O(G, H).$

This can also be recognized as a syllogism, in which ' *F* ' is middle term, ' *G* ' is minor term, and ' *H* ' is major term. It is in fact closely related to (10), in that it is obtained from (10) by merely exchanging ' *F* ' and ' *H* '. There is an obscurity in traditional accounts of the syllogism as to whether (11) counts as the *same* syllogism as (10) or not. Here, we regard them as distinct syllogisms, but as exhibiting the *same pattern*.

Since there are indefinitely many terms, there are indefinitely many distinct syllogisms; but there are only finitely many distinct *patterns* of syllogisms. To see this, let us agree to *fix* the letters of the conclusion as ' *F* ' and ' *H* ' in that order, and use ' *G* ' for the third term: then ' *F* ' is the minor term, ' *G* ' the middle term, and ' *H* ' the major term. Since by definition the order of terms in the conclusion is fixed, there are just four possible ways of permuting the two terms in the two premisses, as the following diagram shows:

	I	II	III	IV
Major premiss:	.. (G, H)	.. (H, G)	.. (G, H)	.. (H, G)
Minor premiss:	.. (F, G)	.. (F, G)	.. (G, F)	.. (G, F)
Conclusion:	.. (F, H)	.. (F, H)	.. (F, H)	.. (F, H)

These four ways are the four *figures* of the syllogism. In Figure I, there are four possible major premisses (A, E, I, or O), four possible minor premisses, and four possible conclusions. Hence there are $4 \times 4 \times 4 = 64$ possible syllogisms. The same calculation holds for the remaining three figures, so that in all we have $4 \times 64 = 256$ possible syllogisms, for the fixed letters '*F*', '*G*', and '*H*'. *Any* syllogism can be obtained from one of these 256 by re-lettering; thus both (10) and (11) are re-letterings of a syllogism in Figure IV (this is most easily seen by inspecting the lay-out of the middle term). Hence there are exactly 256 distinct patterns of syllogism. We may call the 256 actual syllogisms obtained by fixing '*F*', '*G*', and '*H*' in the above manner the *standard instances* of those patterns.

It is of course essential to this calculation that the *order* of terms in the conclusion be fixed by definition. For example,

(12) A (F, H)
      I (H, G)
      ———
      O (F, G)

obtained from (10) by altering the order of '*F*' and '*G*' in the conclusion, is *not* a syllogism since the *second* term '*G*' in the conclusion does not appear in the *first* premiss. If, however, we alter the *order of premisses* as well, to obtain

(13) I(H, G)
      A(F, H)
      ———
      O(F, G),

the result *is* a syllogism ('*H*' still the middle term, '*F*' now minor, and '*G*' major): but, whilst (10) was in the pattern of Figure IV, (13) is in the pattern of Figure I. Thus an old dispute, traces of which occur in quite recent books on traditional logic, as to whether there are three or four distinct figures, can be seen to have arisen out of an uncertainty in the definition of a syllogism.

The main burden of traditional logic is to distinguish, of the 256 possible patterns, which are *valid* and which *invalid*. Two quite separate approaches are used, which yield the same result. One method is to lay down very general principles against which each pattern can in turn be checked. One such principle is that no conclusion follows from two (universal or particular) negative premisses. This principle alone invalidates 16 patterns in each figure, or 64 in all. We shall not state these principles here, but they may be found in the Exercise at the end of this section. The second method, which is Aristotle's own, is to accept as valid certain ' self-evident ' patterns in the first figure and then, using such principles as (1)–(9), to *deduce* the valid patterns of the remaining figures. This method is traditionally known as *reduction to the first figure*, and is said to take two forms, direct and indirect reduction. Roughly speaking, in indirect reduction the valid pattern is deduced by RAA—a contradiction is derived from the supposition that the pattern does *not* hold good, whilst in direct reduction this is not so. Reduction is of great historical interest, as being perhaps the earliest known attempt to derive conclusions systematically from given assumptions. Though Aristotle's presentation is crude and informal by modern standards of rigour, it is possible to follow the outlines of his programme and derive the valid syllogisms as con-clusions, by purely propositional calculus reasoning, from a very small set of syllogistic assumptions. A good account of such a treatment, as of traditional logic in general, may be found in Łukasiewicz [12].

It turns out in fact that there are only 24 valid patterns of syllogism, 6 in each figure. Before tabulating these, however, I wish to relate the theory of the syllogism to the predicate calculus.

A natural question to begin from is: if we translate principles (1)–(9) into the predicate calculus notation, using the translations indicated at the beginning of this section where the four traditional forms were introduced, are the results theorems of the predicate calculus or not? The answer is that some are and some are not; to see why some are not, it will be best to begin with (4), which translates into

(14) $(\exists x)(Fx \mathbin{\&} Gx) \vee (\exists x)(Fx \mathbin{\&} -Gx).$

This is not a theorem of predicate calculus; nor should we wish it to be; for it is easy to prove the following result:

143 $(\exists x)(Fx \mathbin{\&} Gx) \vee (\exists x)(Fx \mathbin{\&} -Gx) \dashv\vdash (\exists x)Fx,$

which shows that (14) is interderivable with the bald assertion that something has F.[1] Hence if (14) were a theorem, it would be a theorem that something had F, for *any* property F, which is clearly absurd. In accepting (4) as a principle of logic, the traditional theory overlooks the possibility that there *may be nothing* with F. Hence I(F, G) and O(F, G) are not strictly subcontrary, as traditional

[1] For proof, compare the proof of the similar (propositional calculus) sequent 45 in Chapter 2, Section 2.

logic maintains, since they will both be false in just the case that there is nothing with F. (It is true neither that some unicorns are fat nor that some unicorns are not fat.)

Let us call a term ' F ' *empty* if there is nothing with F (i.e. if $-(\exists x)Fx$). *Then traditional logic makes the assumption that no term is empty.* Hence in general we can only obtain in predicate calculus the results of the theory of syllogism *on certain existential assumptions.* Thus we obtain (4) in the form (14) on the assumption $(\exists x)Fx$, as 143 reveals. Corresponding to (1) and (2) we *can* prove

144 $\vdash -((x)(Fx \rightarrow Gx) \longleftrightarrow (\exists x)(Fx \And -Gx))$

145 $\vdash -((x)(Fx \rightarrow -Gx) \longleftrightarrow (\exists x)(Fx \And Gx))$

—here no existential assumptions are required. For (3), however, the strongest result obtainable is

146 $(\exists x)Fx \vdash -((x)(Fx \rightarrow Gx) \And (x)(Fx \rightarrow -Gx))$.

It is not hard to see why the existential assumption is required. If nothing has F, then by sequent 133 both $(x)(Fx \rightarrow Gx)$ and $(x)(Fx \rightarrow -Gx)$ will be true (146 can in fact be strengthened to an interderivability). Similarly, for (5) and (6) we obtain at best

147 $(\exists x)Fx \vdash (x)(Fx \rightarrow Gx) \rightarrow (\exists x)(Fx \And Gx)$

148 $(\exists x)Fx \vdash (x)(Fx \rightarrow -Gx) \rightarrow (\exists x)(Fx \And -Gx)$,

for if nothing has F, it will be true that $(x)(Fx \rightarrow Gx)$ and that $(x)(Fx \rightarrow -Gx)$ by 133, yet evidently false that $(\exists x)(Fx \And Gx)$ and that $(\exists x)(Fx \And -Gx)$. (7) and (8) become theorems without existential assumptions, whilst (9) again requires that $(\exists x)Fx$ be assumed.

A square of opposition can be formulated for the predicate calculus, in which the traditional relations do hold, by using in place of the A-, E-, I-, and O-forms the simple quantifier-forms ' $(x)Fx$ ', ' $(x)-Fx$ ', ' $(\exists x)Fx$ ', and ' $(\exists x)-Fx$ '. For we can prove as theorems all of

149 $\vdash -((x)Fx \longleftrightarrow (\exists x)-Fx)$

150 $\vdash -((x)-Fx \longleftrightarrow (\exists x)Fx)$

151 $\vdash -((x)Fx \And (x)-Fx)$

152 $\vdash (\exists x)Fx \text{ v } (\exists x)-Fx$

153 $\vdash (x)Fx \rightarrow (\exists x)Fx$

154 $\vdash (x)-Fx \rightarrow (\exists x)-Fx$

(proofs are elementary). 152 should be compared with (14), the corresponding traditional result. 152, unlike (14), does not entail that something has F, but it does entail that *there is something*. That something either has F or does not; hence 152. Thus 152 reveals clearly the dependence of predicate calculus upon interpretation in *non-empty universes of discourse*. If there were nothing at all, 152 would be false. But the theory of the syllogism makes the far more startling assumption that any *term* is non-empty, that every property has instances.

In order now to exhibit the valid patterns of syllogism, we adopt a convenient shorthand whereby merely the figure and type of proposition appear. Thus ' I EIO ' refers to that pattern in Figure I in which the major premiss is E, the minor I, and conclusion O. The standard instance of this pattern is, therefore,

(15) $E(G, H)$
$\underline{\quad I(F, G) \quad}$
$O(F, H)$.

By the corresponding predicate calculus sequent to a syllogism, we mean the sequent in which the two premisses of the syllogism appear (translated) as assumptions and the conclusion of the syllogism (translated) as conclusion. Thus the sequent corresponding to (15) is

$$(x)(Gx \rightarrow -Hx), (\exists x)(Fx \,\&\, Gx) \vdash (\exists x)(Fx \,\&\, -Hx).$$

The following table gives the 24 valid syllogistic patterns by figure; in deriving the sequents corresponding to the standard instances of these patterns, we require extra existential assumptions in nine cases, as shown in the table:

I AAA	II EAE	† III AAI	§ IV AAI
I AII	II AEE	III AII	IV AEE
I EAE	II AOO	III IAI	† IV EAO
I EIO	II EIO	† III EAO	IV EIO
* I AAI	* II EAO	III EIO	IV IAI
* I EAO	* II AEO	III OAO	* IV AEO

$*$: requires as added assumption $(\exists x)Fx$

† : ,, ,, ,, ,, $(\exists x)Gx$

§ : ,, ,, ,, ,, $(\exists x)Hx$.

The patient reader who has carried out all earlier exercises has already shown the validity of 13 of the corresponding sequents (102, 3.2.2(*a*)–(*d*), 106, 3.3.2(*a*)–(*g*)). He might like now to identify these 13 with 13 in the above table. Existential assumptions are required for exactly those nine valid patterns of syllogism in which a particular conclusion is drawn from two universal premisses, and it is not hard to see why. For if all three terms of the syllogisms are empty, the universal premisses will be true by 133 whilst their conclusions will be false. However, we never require more than one existential assumption to show validity. For example, in the case of IV AAI we can derive in the predicate calculus

155 $(\exists x)Hx, (x)(Hx \rightarrow Gx), (x)(Gx \rightarrow Fx) \vdash (\exists x)(Fx \ \& \ Hx),$

for from $(\exists x)Hx$ and $(x)(Hx \rightarrow Gx)$ it follows that $(\exists x)Gx$, and from $(\exists x)Gx$ and $(x)(Gx \rightarrow Fx)$ it follows that $(\exists x)Fx$, by 105.

From these results the relation between the traditional doctrine of the syllogism and the predicate calculus emerges. The square of opposition principles, the laws of conversion, and the 24 valid patterns of syllogism are all derivable as theorems or sequents of the predicate calculus. Admittedly, in some cases special existential assumptions need to be made. But rather than as a sign of any fundamental discrepancy between the two, this may be viewed as a situation in which predicate calculus helps to make explicit the foundations on which the theory of syllogisms is based. The traditional theory, in fact, is *that fragment of the predicate calculus in which four forms of proposition are selected for special study, it being assumed also that the terms appearing in these forms are not empty*. The predicate calculus is the broader study, at least in the respect that it countenances empty terms; in previous sections, we have seen how it enables us to handle arguments in which propositions appear which are not in any of the four traditional forms; and in the next section, we shall see how it also enables us to deal formally with properties of relations, which lie outside the scope of the theory of the syllogism.

EXERCISE

Let us call the first term of an A, E, I, or O proposition its *subject*, and the second term its *predicate*. Let us also agree to call the predicates of negative propositions and the subjects of universal propositions *distributed*, whilst the predicates of affirmative and the subjects of singular propositions shall be *undistributed*. Using '*d*' for '*distributed*' and '*u*' for 'undistributed', these agreements are shown in the following table:

A (F, G)	E (F, G)
d u	*d d*
I (F, G)	O (F, G)
u u	*u d*

The general principles of the syllogism may now be stated as follows:

A *Rules of Quantity*

1 No term is distributed in the conclusion of a valid syllogism unless it is distributed in the appropriate premiss.

2 The middle term of a valid syllogism is distributed at least once.[1]

B *Rules of Quality*

1 The conclusion of a valid syllogism is negative if and only if one of its premisses is negative.

2 There is no valid syllogism with two negative premisses.

Show informally from these rules:

(*a*) that no valid syllogism contains two particular premisses (Hint: the only possible combinations are II, IO, OI, and OO, of which the last violates B2 and the first, by the table of distribution, violates A2. In the other two cases, by B1 the conclusion must be O or E if the syllogism is to be valid, in either of which case its predicate is distributed, and so by A1 is distributed also in the major premiss: a contradiction follows);

(*b*) that no valid syllogism with a particular premiss has a universal conclusion (Hint: if the conclusion were A, both minor and middle terms would need to be distributed in the premisses, by A1 and A2; and if the conclusion were E, both minor, major, and middle terms would need to be distributed in the premisses, by A1 and A2; use B1 and B2 to show that no permissible combination of premisses allows this);

(*c*) that, of the 256 possible patterns of syllogism, B2 rules out 64 as not valid, the result of (*a*) rules out a further 48, rule B1 rules out a further 72, and the result of (*b*) rules out a further 24 (thus a total of 208 patterns are already shown to be invalid);

[1] To violate this rule is to commit what is sometimes called the fallacy of undistributed middle.

(*d*) that a valid syllogism in Figure I must have an affirmative minor premiss and a universal major premiss;

(*e*) that a valid syllogism in Figure II must have one negative premiss and the major premiss must be universal;

(*f*) that a valid syllogism in Figure III must have a particular conclusion, and, if the conclusion is negative, so is the major premiss.

(In virtue of (*d*)–(*f*), of the 48 patterns remaining for consideration after (*c*), 18 are shown to be valid in Figures I–III and 18 are shown to be invalid: of the remaining 12, 6 can be shown to be valid in Figure IV and the remaining 6 not to be valid by consideration of each case—there are no simple rules for Figure IV.)

5 PROPERTIES OF RELATIONS

If we select a sentence and drop from it a proper name, we obtain a predicate; this predicate, we say, expresses a property. For example, if we drop the proper name ' oxygen ' from ' oxygen is an element ', we obtain the predicate ' . . . is an element ' expressing the property of being an element. Or if we drop ' Mary ' from the sentence ' everyone likes Mary ', we obtain the predicate ' everyone likes . . .' expressing the property of being liked by everyone. In the predicate calculus, predicates become propositional functions in one variable: being an element might be expressed by ' Ex ' in ' x '; being liked by everyone by ' $(y)Lyx$ ' in ' x '.

If we drop two or more proper names from a sentence, we obtain a (*dyadic or polyadic*) *relational expression*, which we say expresses a relation. For example, if we drop ' Brutus ' and ' Caesar ' from ' Brutus killed Caesar ', we obtain the dyadic relational expression '. . . killed . . .', which expresses a certain relation. In the predicate calculus, relational expressions become propositional functions in two or more variables: the relation of killing might be expressed by ' Kxy ' in ' x ' and ' y '.

Propositional functions in any number of variables can be turned into wffs expressing *propositions* in two main ways, which may be combined: we may replace the variables in them by terms, or we may prefix quantifiers. Thus from ' Kxy ' we obtain the wff ' $(x)Kxm$ ' by combining these approaches.

I define neither properties nor relations. These notions are supposed to be understood by examples. The process of explanation has to stop somewhere (though it does not *in fact* have to stop

179

here). For the rest of this section, I shall follow normal logical practice in using ' R ' in place of ' F ' for relations. ' R ' of course counts as a predicate letter, along with ' F ', ' G ', ' H ', . . .; but it will help to emphasize that it is relations with which we are dealing. Only *dyadic* relations (expressed by propositional functions in *two* variables) will be considered.

We shall in fact define various important properties which relations may have, and then study the interconnections between these properties. First, a relation R is said to be *symmetric* if, for any x and y, if R holds between x and y then R holds between y and x; in symbols, R is symmetric if and only if

(1) $(x)(y)(Rxy \rightarrow Ryx)$.

The relation of being the same age as is symmetric; for if a is the same age as b, then b is the same age as a, for arbitrary a and b. The relation of being either the brother or sister of is symmetric; for if a is either the brother or sister of b, then b is either the brother or sister of a. On the other hand, being the brother of is *not* symmetric; for Prince Charles is the brother of Princess Anne, but Princess Anne is not the brother of Prince Charles. In general, to show that a relation R is not symmetric, objects m and n need to be cited such that both Rmn and $-Rnm$. For the negation of (1) is equivalent to (interderivable with)

(2) $(\exists x)(\exists y)(Rxy \mathbin{\&} -Ryx)$.

The relation of being placed next to is symmetric, though the relation of being placed to the right of is not. In view of Theorem 139, the identity relation $=$ is symmetric; (1) in fact becomes 139, if ' R ' is taken as ' $=$ '.

Second, a relation R is *asymmetric* if, for any x and y, if R holds between x and y then R does *not* hold between y and x; in symbols, R is asymmetric if and only if

(3) $(x)(y)(Rxy \rightarrow -Ryx)$.

The relation of being a parent of is asymmetric, since, if a is a parent of b, b is not a parent of a. The numerical relation of being less than is asymmetric, since if a is less than b then b is not less than a. Using ' $<$ ' to express this relation, we have as a truth of mathematics that $(x)(y)(x < y \rightarrow - y < x)$, which satisfies (3) for 'R' as '$<$'.

Loving is not, it seems, an asymmetric relation. For Antony loved Cleopatra, but then also Cleopatra loved Antony. In general, to show that R is not asymmetric, objects m and n need to be cited such that both Rmn and Rnm. For the negation of (3) becomes

(4) $(\exists x)(\exists y)(Rxy \ \& \ Ryx)$.

The mathematical relation of being less than or equal to (in symbols '$\leq$') is not asymmetric. For $2 \leq 2$ and $2 \leq 2$; hence $(\exists x)(\exists y)(x \leq y \ \& \ y \leq x)$—this follows by two steps of EI. (3) and (4) do not require that x and y be distinct objects. We may say that a relation R is *antisymmetric* if, for any *distinct* x and y, if R holds between x and y, then R does not hold between y and x: in symbols, R is antisymmetric if and only if

(5) $(x)(y)(x \neq y \ \& \ Rxy \rightarrow - Ryx)$.[1]

Then $\leq$, though not asymmetric, is antisymmetric, since if a and b are *distinct* numbers such that $a \leq b$ it does follow that $- b \leq a$. An equivalent formulation of the definition of antisymmetry is

(6) $(x)(y)(Rxy \ \& \ Ryx \rightarrow x = y)$

—a relation R is antisymmetric if and only if for any x and y such that R holds both between x and y and between y and x, x is identical with y.

In studying the interconnections between the properties of relations so far defined, the first point to notice is that a relation will not usually be both symmetric and asymmetric. But it will be in the extreme case that the relation fails to hold between anything. For it is easy to prove

156 $- (\exists x)(\exists y) \ Rxy \vdash (x)(y)(Rxy \rightarrow Ryx)$

157 $- (\exists x)(\exists y) \ Rxy \vdash (x)(y)(Rxy \rightarrow - Ryx)$

(these are 'relational analogues' of 133). For example, the relation of being a female brother of holds between no objects, and so is both symmetric and asymmetric (and for that matter antisymmetric). On the other hand, if a relation holds *at all*, it cannot be both symmetric and asymmetric; we may readily prove

[1] We use '$x \neq y$' as a convenient shorthand for '$- (x = y)$'.

181

158 $(\exists x)(\exists y)Rxy \vdash -((x)(y)(Rxy \rightarrow Ryx) \mathbin{\&} (x)(y)(Rxy \rightarrow -Ryx))$.

The need for the existential assumption in 158 corresponds closely to the similar need in the case of certain principles of the theory of the syllogism, as we saw in the last section.

Though a relation which holds at all cannot be *both* symmetric and asymmetric, it may be *neither*. For example loving, which we have seen to be not asymmetric, is also not symmetric. Any example of unreturned affection will satisfy (2). We may say that a relation R is *non-symmetric* if it is neither symmetric nor asymmetric; a symbolic definition is obtained simply by forming the conjunction of (2) and (4). We can then prove absolutely that *any* relation is either symmetric or asymmetric or non-symmetric; and we can prove that, if a relation holds at all, then it is not more than one of these three. (If a relation fails to hold, then, though it is symmetric and asymmetric, it is not non-symmetric.)

Any asymmetric relation is antisymmetric. It is easy to prove

159 $(x)(y)(Rxy \rightarrow -Ryx) \vdash (x)(y)(x \neq y \mathbin{\&} Rxy \rightarrow -Ryx)$

—given the condition for asymmetry of R, the condition for R's antisymmetry follows. The converse, of course, does not hold; for $\leq$ is antisymmetric but not asymmetric. $\leq$ is not symmetric ($3 \leq 4$, but it is not the case that $4 \leq 3$), so that it serves as an example of an antisymmetric relation which is non-symmetric. An antisymmetric relation may also be symmetric. Oddly enough, $=$ is an example, for we can easily prove that $(x)(y)(x \neq y \mathbin{\&} x = y \rightarrow y \neq x)$.

A quite distinct property which many relations possess is that of transitivity. A relation R is *transitive* if, for any x, y, and z, if R holds between x and y and between y and z, then it holds between x and z. Thus R is transitive if and only if

$(7)\quad (x)(y)(z)(Rxy \mathbin{\&} Ryz \rightarrow Rxz)$.

Being of the same age as is transitive, as well as symmetric; for if a is of the same age as b, and b of the same age as c, then a is of the same age as c. Identity is transitive in view of 140. The relation $<$ is clearly transitive, though asymmetric. $\leq$ is also transitive, though non-symmetric. Hence transitivity cuts right across the previous distinctions. Being not identical with (different from) is not, however, transitive, though often supposed to be. For example, $10 \neq 11$, and $11 \neq 7 + 3$, but $10 = 7 + 3$. (7) does not stipulate that x, y, and z be distinct objects.

R is *intransitive* if and only if

$\qquad$ (8) $(x)(y)(z)(Rxy \ \& \ Ryz \rightarrow -Rxz)$.

Being a parent of is intransitive (and asymmetric); for if a is a parent of b, and b a parent of c, a is not a parent (but a grandparent) of c. The numerical relation of differing by 1 from is intransitive; for if a differs by 1 from b and b differs by 1 from c, either $a = c$ or a differs by 2 from c, and in either case a does not differ by 1 from c. This relation is also symmetric. There are also intransitive non-symmetric relations; try to think of one.

A relation R is *not* transitive if and only if

$\qquad$ (9) $(\exists x)(\exists y)(\exists z)((Rxy \ \& \ Ryz) \ \& \ -Rxz)$.

(9) can be obtained by suitably transforming the negation of (7). In a similar way, R is not intransitive if and only if

$\qquad$ (10) $(\exists x)(\exists y)(\exists z)((Rxy \ \& \ Ryz) \ \& \ Rxz)$.

In parallel to the definition of non-symmetry we may define R to be *non-transitive* in case R is neither transitive nor intransitive; a symbolic formulation results from conjoining (9) and (10). As in the case of the earlier relations, it is easy to show that, if a relation R fails to hold at all, then R is both transitive and intransitive. Further, all relations are either transitive, intransitive, or non-transitive.

Three more properties of relations are important, but they are not independent of the earlier properties, as we shall see. A relation R is *reflexive* if, for any x, R holds between x and itself, x; thus R is reflexive if and only if

$\qquad$ (11) $(x)Rxx$.

In view of 138, $=$ is reflexive; so are the relations of being the same age as, being the same height as, having the same-coloured eyes as, and having the same parents as. For everyone is his own age and height, etc. R is *irreflexive*, on the other hand, if for any x, it is not the case that R holds between x and x; thus R is irreflexive if and only if

$\qquad$ (12) $(x)-Rxx$.

Being different from is an irreflexive relation. A relation R may be neither reflexive nor irreflexive, and will be just in case

(13) $(\exists x)Rxx \ \& \ (\exists x) - Rxx.$

We then say that R is *non-reflexive*. Presumably loving is non-reflexive; there are those who love themselves (hence it is not irreflexive), and there are those who do not (hence it is not reflexive).

Every relation is either reflexive, irreflexive, or non-reflexive, and not more than one of these three. This is true whether the relation holds at all or not at all; for if the relation holds not at all, it is irreflexive, but neither reflexive nor non-reflexive. We can easily prove

160 $-(\exists x)(\exists y)Rxy \vdash (x) - Rxx$

161 $-(\exists x)(\exists y)Rxy \vdash -(x)Rxx$

162 $-(\exists x)(\exists y)Rxy \vdash -((\exists x)Rxx \ \& \ (\exists x) - Rxx).$

This difference between the present group of properties and the earlier groups clustering round symmetry and transitivity arises essentially because the three latest definitions are not *conditional* in form, and so cannot turn out trivially true because their antecedents are always false, as happened in the earlier cases. In fact, 161 reflects our assumption that no universe is empty; if a relation is reflexive, i.e. if $(x)Rxx$, by 104 it follows that $(\exists x)Rxx$, so that R holds at least in some case.

Of the various interconnections that exist between the 10 defined properties of relations, three of the most important are proved in the following three sequents. First, *all asymmetric relations are irreflexive*, as we see from the sequent

163 $(x)(y)(Rxy \rightarrow -Ryx) \vdash (x) - Rxx$

1	(1)	$(x)(y)(Rxy \rightarrow -Ryx)$	A
1	(2)	$(y)(Ray \rightarrow -Rya)$	1 UE
1	(3)	$Raa \rightarrow -Raa$	2 UE
1	(4)	$-Raa$	3 SI(S) 23
1	(5)	$(x) - Rxx$	4 UI

We assume the asymmetry condition at line (1), and derive the irreflexivity condition at line (5). The trick in the proof lies in the double application of UE using the *same* arbitrary name ' *a* ' both times. The reader should satisfy himself that he understands the soundness of the step from (2) to (3). The step from (3) to (4) is at the level of propositional calculus.

Secondly, *all irreflexive and transitive relations are asymmetric.* In 164, we obtain the condition for asymmetry as a conclusion from the conditions for irreflexivity and transitivity.

164 $(x) - Rxx, (x)(y)(z)(Rxy \ \& \ Ryz \rightarrow Rxz) \vdash (x)(y)(Rxy \rightarrow - Ryx)$

1	(1) $(x) - Rxx$	A
2	(2) $(x)(y)(z)(Rxy \ \& \ Ryz \rightarrow Rxz)$	A
2	(3) $(y)(z)(Ray \ \& \ Ryz \rightarrow Raz)$	2 UE
2	(4) $(z)(Rab \ \& \ Rbz \rightarrow Raz)$	3 UE
2	(5) $Rab \ \& \ Rba \rightarrow Raa$	4 UE
1	(6) $- Raa$	1 UE
1,2	(7) $- (Rab \ \& \ Rba)$	5,6 MTT
1,2	(8) $- Rab \ \lor \ - Rba$	7 SI(S) 1.5.1(g)
1,2	(9) $Rab \rightarrow - Rba$	8 SI(S) 48
1,2	(10) $(y)(Ray \rightarrow - Rya)$	9 UI
1,2	(11) $(x)(y)(Rxy \rightarrow - Ryx)$	10 UI

The trick here is the step of UE from (4) to (5), in which ' *z* ' is eliminated in favour of the already present ' *a* '. This yields a consequent *Raa*, whose negation we can obtain from (1) (line (6)), and so paves the way for the propositional calculus reasoning leading to (9). Since (1) and (2) lack both ' *a* ' and ' *b* ', the steps of UI at lines (10) and (11) are justified.

Thirdly, *all intransitive relations are irreflexive.* The proof of the sequent containing this information is basically like that of 163.

165 $(x)(y)(z)(Rxy \ \& \ Ryz \rightarrow - Rxz) \vdash (x) - Rxx$

1	(1) $(x)(y)(z)(Rxy \ \& \ Ryz \rightarrow - Rxz)$	A
1	(2) $(y)(z)(Ray \ \& \ Ryz \rightarrow - Raz)$	1 UE
1	(3) $(z)(Raa \ \& \ Raz \rightarrow - Raz)$	2 UE
1	(4) $Raa \ \& \ Raa \rightarrow - - Raa$	3 UE
5	(5) Raa	A

5	(6) *Raa* & *Raa*	5,5 &I
1,5	(7) $-Raa$	4,6 MPP
1,5	(8) *Raa* & $-Raa$	5,7 &I
1	(9) $-Raa$	5,8 RAA
1	(10) $(x) -Rxx$	9 UI

From these three principles various conclusions may be drawn about what combinations of properties relations cannot have. One example will suffice. If a relation holds at all, then it cannot be irreflexive and transitive and symmetric. For if it is irreflexive and transitive, by 164 it is asymmetric. And if it holds at all, as we saw earlier, it is not both symmetric and asymmetric. It is important to notice that this conclusion essentially depends on the assumption that the relation holds; for if $- (\exists x)(\exists y)Rxy$, then R is both irreflexive and transitive and symmetric.

The study of relations can be pursued a great deal further than we are able to take it in this book. But it is best carried out within the framework of a broader logical discipline than the predicate calculus. The discipline at present favoured for this undertaking is *the theory of classes*, the study of classes of objects and the relation of membership between an object in a class and the class itself. In terms of classes, the notion of a relation can itself be defined— relations turn out in fact to be a special kind of class. Class theory is indeed so powerful a logical system that all *mathematical* concepts can be defined in it: natural numbers, rational numbers, real numbers can be handled as special kinds of classes. This possibility has led certain philosophers to advance the thesis that all mathematics can be reduced to logic—a thesis which again it is outside the scope of this book to discuss. However, a study of class theory may be regarded as the next step which the reader who wishes to go further should take; an excellent introduction is Suppes [25], and Appendix B of this book outlines the early stages of this theory.

Although we have not developed a notation for expressing them exactly, there are certain notions so commonly employed in connection with relations that it is worth, for reference purposes, giving here an informal account of them. A formal account is forthcoming within class theory. By the *domain* of a relation *R* we understand the class of objects which bear the relation *R* to something. Thus the domain of the relation of loving is the class of all lovers, the domain of the relation being a parent of is the class of all fathers and mothers. On the other hand, by the *converse domain* (sometimes called the *range*) of a relation R we understand the class of objects such that something bears the relation R to them. Thus the converse domain of the relation of loving is the class of all loved things, the converse domain of the relation of being a parent of is the class of all things with parents, which, if we are to believe the Bible, does not include everyone. The *field* of a relation R is the class of things either in the domain or in the converse domain of R: the class of things, in fact, which either bear relation R to something or have relation R borne to them by something. Thus the field of the loving relation is the class of things loving or loved—a class which, presumably, does not have Scrooge as a member. A relation which holds of nothing has an *empty* field—a field which has no members.

A slightly more formal presentation of these ideas would be the following. a belongs to the domain of R if and only if

(14) $(\exists x)Rax$;

a belongs to the converse domain of R if and only if

(15) $(\exists x)Rxa$;

a belongs to the field of R if and only if

(16) $(\exists x)Rax \lor (\exists x)Rxa$.

We shall not pursue the formal development of logic beyond this point. The bibliography, however, contains suggestions for further reading.

EXERCISES

1 (*a*) Prove the interderivability of (5) and (6) in the text.
   (*b*) Prove the validity of sequents 156–162 in the text.

2 Let us call a relation R *serial* if $(x)(\exists y)Rxy$, i.e. if everything bears the relation R to something. Thus being less than (' $<$ ') is serial in the universe of natural numbers, since for any number there is a number than which it is less (there is no greatest natural number).

   (*a*) Prove that $=$ is serial: i.e. prove $\vdash (x)(\exists y)(x = y)$.

   (*b*) Prove that if a relation R is serial then it is not empty: i.e. prove $\vdash (x)(\exists y)Rxy \rightarrow (\exists x)(\exists y)Rxy$.

   (*c*) Prove that all serial, transitive, and symmetric relations are reflexive.

187

3 Given that a relation is transitive, show that it is irreflexive if and only if it is asymmetric.

4 The relation of being a brother or sister of seems intuitively both transitive and irreflexive, whence by 164 it would follow that it was asymmetric, which it clearly is not. How do you explain this paradox?

5 Given that R is both symmetric and antisymmetric, prove that no two distinct things stand in the relation R to each other (i.e. $(x)(y)(x \neq y \to -Rxy)$) and further that R is transitive. Hence show that if R is reflexive, symmetric, and antisymmetric, then $(x)(y)(Rxy \leftrightarrow x = y)$. (This effectively shows that identity is the only reflexive, symmetric, and antisymmetric relation.)

6 Show that no relation can be:

   (a) intransitive and reflexive;

   (b) asymmetrical and non-reflexive;

   (c) transitive, reflexive, and asymmetric;

   (d) transitive, non-symmetric, and irreflexive.

7 (a) From 156–158 it can readily be seen to follow that relation R is both symmetric and asymmetric if and only if $-(\exists x)(\exists y)Rxy$. Prove correspondingly that a relation R is both transitive and intransitive if and only if $-(\exists x)(\exists y)(\exists z)(Rxy \ \& \ Ryz)$.

   (b) Hence prove that if R is both transitive and intransitive, then R is asymmetric.

APPENDIX A

Normal Forms

It is customary to include in elementary logic courses a treatment of normal forms. Normal forms have a certain interest in connection with the truth-table method, since they provide an independent test as to whether a wff is tautologous, contingent, or inconsistent; and they are also used in certain proofs of the completeness of the propositional calculus (see, e.g., Basson and O'Connor [1]). The completeness proof given in this book (Chapter 2, Section 5), however, did not rely on normal forms, so that I have relegated an account of them to an appendix. What follows presupposes the terminology of Chapter 2, Section 3.

We begin by defining normal forms. First, by an *atom* I understand *either a propositional variable or a negated propositional variable*. Thus

$$`P`, `-P`, `Q`, `-R`, `S`$$

are all atoms, though

$$`--Q`$$

is not one. Let $A_1, \ldots, A_n$ be a list of n atoms, where n is greater than or equal to 1. Then by an *elementary disjunction* (e.d.) I understand a formula of the form

$$(A_1 \vee A_2 \vee \ldots \vee A_n).$$

Thus any list of atoms linked by ' v 's counts as an elementary disjunction: for example,

$$`(P \vee Q)`$$
$$`(-P \vee Q \vee -R)`$$
$$`(-Q \vee P \vee Q \vee -S)`$$
$$`(P \vee P \vee -Q)`$$

are all elementary disjunctions. In the limiting case where $n = 1$, a single atom standing alone counts as an elementary disjunction also.

The same variable may appear both negated and non-negated in an e.d., as in the third example, and the same atom may appear more than once, as in the fourth.

Correspondingly, by an *elementary conjunction* (e.c.) I understand a formula of the form

$$(A_1 \ \& \ A_2 \ \& \ldots \& \ A_n)$$

for atoms $A_1, \ldots, A_n$, where n is greater than or equal to 1. Any e.d. becomes an e.c. if all the ' v 's in it are changed to ' & 's, and vice versa. Again, in the limiting case, an atom standing alone counts as an e.c.

By a *conjunctive normal form* (C.N.F.) I understand a formula of the form

$$A_1 \ \& \ A_2 \ \& \ldots \& \ A_n,$$

where $A_1, \ldots, A_n$ are elementary *disjunctions*, for n greater than or equal to 1. Thus a C.N.F. is a string of e.d.'s linked by ' & '. For example,

$$\text{`} (P \lor Q) \ \& \ (P \lor -Q \lor R) \ \& \ (-R \lor S) \text{'}$$
$$\text{`} (P \lor -P \lor Q) \ \& \ -S \text{'}$$
$$\text{`} (P \lor -Q) \ \& \ (P \lor -Q) \text{'}$$

are all C.N.F.'s, the second of which has as its second e.d. a single atom and the third of which has as its two conjuncts the same e.d. In the limiting case where $n = 1$, a single e.d. standing alone counts as a C.N.F., so that '$(P \lor Q)$' or even 'P' alone counts as a C.N.F.

Correspondingly, by a *disjunctive normal form* (D.N.F.) I understand a formula of the form

$$A_1 \lor A_2 \lor \ldots \lor A_n,$$

where $A_1, \ldots, A_n$ are elementary *conjunctions*, for n greater than or equal to 1. Thus a D.N.F. is a string of e.c.'s linked by ' v '. Any C.N.F. becomes a D.N.F. if all ' v 's are changed to ' & 's and all ' & 's to ' v 's, and vice versa. In the limiting case where $n = 1$, a single e.c. standing alone counts as a D.N.F.

A consequence of these definitions is that all e.c.'s and e.d.'s are both C.N.F.'s and D.N.F.'s. For example, consider

$$\text{`} P \ \& \ -Q \ \& \ R \text{'},$$

an e.c. This is a C.N.F. whose three e.d.'s are each of them the limiting case of a single atom. It is also a D.N.F. in the limiting case where the number of e.c.'s is 1. As an extreme limiting case, a single atom is both a C.N.F. and a D.N.F.

I now describe a procedure for *reducing* any wff of the propositional calculus to a C.N.F. and to a D.N.F. This will consist in finding, for any wff, a C.N.F. and a D.N.F. *equivalent* to it in an appropriate sense. The steps of the procedure will all be of the same kind, namely the replacement of a part or the whole of a given formula by an equivalent formula. To this end, we note the following biconditionals, which we shall need in our work:

(1) $P \& P \leftrightarrow P$

(2) $P \vee P \leftrightarrow P$

(3) $P \& Q \leftrightarrow Q \& P$

(4) $P \vee Q \leftrightarrow Q \vee P$

(5) $P \& (Q \& R) \leftrightarrow (P \& Q) \& R$

(6) $P \vee (Q \vee R) \leftrightarrow (P \vee Q) \vee R$

(7) $--P \leftrightarrow P$

(8) $P \rightarrow Q \leftrightarrow -P \vee Q$

(9) $(P \leftrightarrow Q) \leftrightarrow (P \rightarrow Q) \& (Q \rightarrow P)$

(10) $-(P \& Q) \leftrightarrow -P \vee -Q$

(11) $-(P \vee Q) \leftrightarrow -P \& -Q$

(12) $P \vee (Q \& R) \leftrightarrow (P \vee Q) \& (P \vee R)$

(13) $P \& (Q \vee R) \leftrightarrow (P \& Q) \vee (P \& R)$.

All of (1)–(13) are provable as theorems of the propositional calculus, and are also tautologies by a truth-table test. (Most should be familiar from exercises and results in the text.) (1) and (2) are sometimes called the *laws of idempotence* for ' & ' and ' v '. (3) and (4) are called the *commutative laws* for ' & ' and ' v '. (5) and (6) are called the *associative laws* for ' & ' and ' v '. (7) is the *law of double negation*. (10) and (11) are forms of *de Morgan's laws*, and (12) and (13) are the *distributive laws*.

It is in virtue of the associative laws (5) and (6) that, for the purposes of truth-table computation, we permit ourselves to write 'complex conjunctions' or 'complex disjunctions', such as '$P \& Q \& -R$'

or ' $-P \lor Q \lor R$ ', without inner brackets. Strictly, such expressions are not wffs (see the definition in Chapter 2, Section 1), and accordingly C.N.F.'s and D.N.F.'s will not in general be well-formed. But, as far as a truth-table evaluation is concerned, it makes no difference how brackets are inserted. We may compare the situation in arithmetic with respect to ' $+$ ', which is also associative: $(x + y) + z = x + (y + z)$, so that we may write ' $5 + 3 + 2$ ' safely without brackets. By contrast, ' $-$ ' is *not* associative: $5 - (3 - 2) = 4$, whilst $(5 - 3) - 2 = 0$, so that ' $5 - 3 - 2$ ' is dangerously ambiguous. Similarly, in logic ' $\rightarrow$ ' is not associative: ' $P \rightarrow (Q \rightarrow R) \leftrightarrow (P \rightarrow Q) \rightarrow R$ ' is not a tautology, as the reader should confirm. (Is ' $\leftrightarrow$ ' associative or not?)

Similarly, in virtue of the commutative laws (3) and (4), the order in which the conjuncts or disjuncts of a complex conjunction or disjunction appear will not affect the truth-table evaluation. Again, ' $\&$ ' and ' $\lor$ ' resemble ' $+$ ' in arithmetic, for which we have $x + y = y + x$; by contrast, ' $-$ ' is not commutative ($5 - 3 = 2$, whilst $3 - 5 = -2$), nor is ' $\rightarrow$ ' in logic, since $P \rightarrow Q \leftrightarrow Q \rightarrow P$ is not a tautology. (Is ' $\leftrightarrow$ ' commutative?)

Further, in virtue of the idempotence laws (1) and (2), we may safely drop reduplicated conjuncts or disjuncts in a complex conjunction or disjunction without affecting the truth-table evaluation. In this respect, ' $\&$ ' and ' $\lor$ ' *differ* from ' $+$ ', since we clearly do not have as an arithmetical truth $x + x = x$. ' $\rightarrow$ ' is not idempotent either. (Is ' $\leftrightarrow$ '?)

In reduction to normal forms, we are seeking, for a given wff, a normal form which will be *equivalent to the wff under a truth-table test*: i.e. for any given assignment of truth-values to the variables of the wff, the normal form shall have the same truth-value as the original wff, so that the biconditional of the wff and its normal form will be a tautology (compare the definition of equivalence in Chapter 3, Section 3). Hence, in virtue of (1)–(6), in the search for a normal form we allow ourselves to take brackets out of, rearrange the order of items in, and delete reduplicated items from, complex conjunctions and disjunctions, whenever such moves suit our purpose. Further, in virtue of (7), we shall allow ourselves freely to drop ' $--$ ' whenever we wish, since such a move cannot affect the truth-table evaluation either.

In fact, all our moves in the transformation of a wff into a normal form are of the same kind: we replace a part (sometimes the whole) of a given formula by some formula equivalent to it under a truth-table test. And the only equivalences we use are (1)–(13) or sub-stitution-instances of them. It should be clear that no move of this kind will affect the truth-value evaluation, so that the normal form which emerges at the end of the process will be equivalent to the given wff, as is desired. But for the record we state, though do not prove, the principle of replacement which guides our work:

(R) If A is equivalent to B, and D results from C by replacing some occurrence of A in C by B, then C is equivalent to D, for any formulae A, B, C, D.

It is in fact this principle, together with (1)–(7), that justifies the moves we have already agreed to make (dropping of ' — — ', re-arrangement of the order of conjuncts in a conjunction, etc.).

The reductions of a wff to a C.N.F. and a D.N.F. may conveniently be divided into two stages, the first of which is common to both reductions.

Stage I. The objective here is to obtain a formula with the following three properties: (i) ' — — ' nowhere appears; (ii) the only connectives that appear are ' — ', ' & ', or ' v '; (iii) ' — ' appears only before propositional variables, nowhere before a bracket (in fact, of course, (iii) implies (i)).

As to (i), this requirement is simply satisfied by dropping ' — — ' in virtue of (7). Our first step, therefore, will be to eliminate any occurrences of ' → ' or ' ↔ ' in the given wff. This can be effected in virtue of (8) and (9), which permit us to replace any conditional by a disjunction the first disjunct of which is the negation of its antecedent and the second disjunct of which is its consequent, and any biconditional by a certain conjunction of two conditionals, each of which in turn can be eliminated in the manner just described.

When this step is complete, we shall have a formula satisfying condition (ii). However, it may be the case that ' — ' appears in this formula outside a bracket, so that (iii) is not satisfied. Still, the main connective inside the bracket can now only be either ' & ' or ' v ', so that the formula has as a part either something of the form

$$-(\ldots \& \ldots)$$

or something of the form

$$-(\ldots v \ldots).$$

If the first case arises, we can apply (10) and replace the part by a disjunction whose disjuncts are the negations of the original conjuncts in the bracket; if the second case arises, we can similarly apply (11) and replace the negated disjunction by a conjunction with negated conjuncts. As a result of these moves, it may still be the case that ' — ' appears before a bracket. But it should be clear that by *repeating* moves of this kind, we shall eventually work the ' — 's into the formula in such a way that they appear, if at all, only before variables, so that sooner or later condition (iii) will also be met, and Stage I will be complete.

The procedures for C.N.F. and for D.N.F. diverge after this point. Accordingly, we describe each separately.

Stage II (a) (continuation for C.N.F.). If, at the end of Stage I, we have not already obtained a C.N.F., this can be for one reason only, as the definition of C.N.F. makes plain: at one or more places there must be an occurrence of ' & ' which is subordinate to some occurrence of ' v '. For if this were not so, given conditions (i)–(iii) of Stage I, we should in fact have a C.N.F. Hence (though this may involve a rearrangement of disjuncts) some part of the given formula is of the form

$$\ldots v (\ldots \& \ldots).$$

By using (12), this can be replaced by a formula of the form

$$(\ldots v \ldots) \& (\ldots v \ldots),$$

in which the ' v ' has moved into subordinate position and the ' & ' into subordinating position. Of course, it may be that this new ' & ' is still subordinate to some other occurrence of ' v '; but in that case we can reapply the same procedure. Eventually, using only the equivalence (12) or its substitution-instances, we can bring all the ' & 's into subordinating positions and relegate all ' v 's into subordinate positions. The result will be a C.N.F. (How can we be sure that all the conjuncts of the C.N.F. will be *elementary* disjunctions?)

Stage II (b) (continuation for D.N.F.). By entirely similar considerations, we can see that, if the result of Stage I is not already a D.N.F., there must be in it at least one occurrence of ' v ' subordinate to an ' & '. Hence some part of it, perhaps after rearrangement of conjuncts, is of the form

$$\ldots \& (\ldots v \ldots).$$

By using (13), this can be replaced by a formula of the form

$$(\ldots \& \ldots) \vee (\ldots \& \ldots),$$

and, by repeating steps of this kind, we can bring all ' v 's into subordinating positions and all ' & 's into subordinate positions. The result will be a D.N.F.

Let us illustrate these procedures for a wff of medium complexity

(i) $S \rightarrowtail -((P \rightarrowtail Q) \rightarrowtail R)$.

Embarking on Stage I, we apply (8) to transform the three con-ditionals into disjunctions. This yields

(ii) $-S \vee -(-(-P \vee Q) \vee R)$.

This satisfies the second condition of Stage I, but there are still two ' — 's outside brackets. We apply (11) to the second disjunct of (ii) to obtain

(iii) $-S \vee (--(-P \vee Q) \& -R)$,

or, dropping a double negation,

(iv) $-S \vee ((-P \vee Q) \& -R)$.

Stage I is now clearly complete, but the result is neither a C.N.F. nor a D.N.F. We set out, therefore, on Stage II (*a*). Observe that, in (iv), the sole occurrence of ' & ' is subordinate to the main connective ' v '. We may, therefore, apply (12) to (iv) as a whole, and obtain

(v) $(-S \vee (-P \vee Q)) \& (-S \vee -R)$.

Within the first conjunct of (v) the brackets are needless by (6), and we have as a C.N.F.

(vi) $(-S \vee -P \vee Q) \& (-S \vee -R)$.

We now revert to (iv), and start Stage II (*b*). The first disjunct ' —S ' of (iv) will do, of course, as an e.c. for our desired D.N.F., and we need only concentrate on the ' v ' which is subordinate ' & '. Rearranging the conjuncts, we obtain

(vii) $-S \vee (-R \& (-P \vee Q))$,

whence, applying (13) to the second disjunct,

(viii) $-S \vee ((-R \& -P) \vee (-R \& Q))$.

This is in effect a D.N.F., and we may drop a pair of brackets to obtain

(ix) $-S \vee (-R \& -P) \vee (-R \& Q)$.

8

Appendix A

In actual practice, Stage II is usually the more formidable, though the principles involved are very simple. This is because an application of the distributive laws doubles the number of brackets, and the resulting formula may be almost twice the length of the original. Students who do normal form work must not be daunted by this, and also need to pay very close attention to the bracketing of their formulae. It should also be remembered that, in Stage II, there is often a choice as to where one begins on subordinate occurrences of ' v 's or ' & 's. As a result, different C.N.F.'s and D.N.F.'s for the same original wff may be obtained—there is no *unique* C.N.F. or D.N.F. for a given formula.

In order to put C.N.F.'s and D.N.F.'s to some use, we first state some obvious equivalences. Let T be *any* tautology, and I *any* inconsistency; then the following are tautologous:

(14) $T \vee P \longleftrightarrow T$

(15) $T \& P \longleftrightarrow P$

(16) $I \vee P \longleftrightarrow P$

(17) $I \& P \longleftrightarrow I.$

By (14), a disjunction with a tautologous disjunct is itself a tautology, and by (17) a conjunction with an inconsistent conjunct is itself an inconsistency. By (15), a conjunction with a tautologous conjunct is equivalent to the other conjunct (and so the former may be deleted as far as a truth-table evaluation goes), and by (16) a disjunction with an inconsistent disjunct is equivalent to the other disjunct (and so the former may be deleted as far as a truth-table evaluation goes).

It follows from (15) that (α) a complex conjunction is tautologous if and only if each of its conjuncts is tautologous, and from (16) that (β) a complex disjunction is inconsistent if and only if each of its disjuncts is inconsistent. From (α) we may infer that an *elementary* conjunction can never be tautologous, for no atom can be tautologous; similarly, no *elementary* disjunction can be inconsistent. However, we can establish

(γ) An e.d. is tautologous if and only if it has among constituent atoms a propositional variable and the negation of the same variable.

For suppose an e.d. does contain a variable, say 'P', and also '$-P$', the negation of that variable. Then, by rearranging the atoms if necessary, we can bring it into the form

$$P \vee -P \vee \ldots,$$

whence, by (14), it is tautologous. Conversely, if it lacks as atoms any variable together with the negation of the same variable, we can find an assignment of truth-values that makes each atom false (namely, for variables that appear negated, the value T, and for variables that appear non-negated, the value F), and so the whole e.d. false. Similarly, using (17), we can show

(δ) An e.c. is inconsistent if and only if it has among its constituent atoms a propositional variable and the negation of the same variable.

From (α) and (γ) we infer that a C.N.F. is tautologous if and only if all its e.d.'s are tautologous, i.e. if and only if *every e.d. in it has amongst its constituent atoms a propositional variable and the negation of the same variable.* That is to say, we can 'read off' from a C.N.F. of a wff whether it is tautologous or not. For example, from (vi) above we can infer that (i) is not tautologous, since (vi) has at least one e.d. lacking a variable and the negation of the same variable (in fact *both* e.d.'s are like that in this case).

Similarly, from (β) and (δ) we infer that a D.N.F. is inconsistent if and only if all its e.c.'s are inconsistent, i.e. if and only if *every e.c. in it has amongst its constituent atoms a propositional variable and the negation of the same variable.* For example, from (ix) above we infer that (i) is not inconsistent, since (ix) has at least one e.c. lacking a variable and the negation of the same variable. In fact, we can now conclude that (i), being neither tautologous nor inconsistent, is contingent. Considering the second conjunct of (vi), we see that the assignment $S = $ T, $R = $ T makes (i) false; and considering the first disjunct of (ix), we see that the assignment $S = $ F makes (i) true. In general, for any wff, we can tell very simply from any C.N.F. equivalent to it whether it is tautologous or not and from any D.N.F. equivalent to it whether it is inconsistent or not, so that reduction to C.N.F. and D.N.F. provides a test as to whether a wff is tautologous, contingent, or inconsistent which is independent of the truth-table test. Reduction to normal form may

197

save us the labour of a truth-table test—at the cost of labour of a different kind.

A certain interest attaches to normal forms of a special kind, which are sometimes called *canonical*. A *canonical conjunctive normal form* (C.C.N.F.) is a C.N.F. in which every propositional variable that occurs (negated or unnegated) in some e.d. occurs (negated or unnegated) in all e.d.'s in the C.N.F. Correspondingly, a *canonical disjunctive normal form* (C.D.N.F.) is a D.N.F. in which every propositional variable that occurs (negated or unnegated) in some e.c. occurs (negated or unnegated) in all e.c.'s in the D.N.F. For example (vi) above is not canonical, since ' R ' appears in the second e.d. but not in the first, and (ix) is not canonical, since ' S ' appears in neither the second nor the third e.c. On the other hand

$$' (P \lor -Q \lor -R) \& (Q \lor -P \lor R) '$$

is a C.C.N.F., and the interchange of ' & ' and ' v ' in it yields a C.D.N.F.

In order to obtain canonical normal forms from non-canonical ones, we note first the two equivalences

(18) $P \longleftrightarrow (P \lor Q) \& (P \lor -Q)$

(19) $P \longleftrightarrow (P \& Q) \lor (P \& -Q)$.

If, now, a given C.N.F. is not a C.C.N.F., there must be some variable occurring in some e.d. in the given form which does not occur in all e.d.'s in the form. We may use (18) to replace any e.d. lacking a given variable by a pair of e.d.'s each containing that variable as well as the other atoms of the given e.d.; in one member of the pair it appears unnegated, and negated in the other. For example, wishing to transform (vi) into a C.C.N.F., we should replace the first conjunct by

$$' (-S \lor -P \lor Q \lor R) \& (-S \lor -P \lor Q \lor -R) ',$$

to which it is equivalent by (18), thus obtaining two e.d.'s in which all four variables of (vi) appear. (There would then remain the task of transforming the second conjunct of (vi) into e.d.'s containing all four variables.) Thus repeated use of (18) will transform a C.N.F. that is not canonical into an equivalent C.C.N.F. In an entirely similar way, we can, using (19), transform any D.N.F. that is not canonical into a C.D.N.F.

Given a canonical normal form, we may simplify it by (i) deleting any repetitions of atoms occurring in any e.d. or e.c. in the form, (ii) deleting any repetitions of e.d.'s or e.c.'s in the form, (iii) in the case of a C.N.F., deleting any tautologous e.d.'s in virtue of (15), and, in the case of a D.N.F., deleting any inconsistent e.c.'s in virtue of (16). In virtue of (iii), the normal form may vanish altogether, and will do if the C.N.F. is tautologous or the D.N.F. inconsistent. In that case, we agree to write ' P v $-P$ ' and ' P & $-P$ ' respectively. Let us call the result of these manoeuvres a *distinguished* (conjunctive or disjunctive) normal form. Then it can be shown that, for each wff, its distinguished (conjunctive or disjunctive) normal form is *unique*, apart from variations in the order of atoms in e.d.'s or e.c.'s and in the order of the e.d.'s or e.c.'s themselves. Moreover, these forms bear a close relation to the truth-table for the given wff, in that the truth-table can be read off from either of them and they can be read off from it.

This is perhaps best shown by an example. Let us suppose a wff A contains the three variables ' P ', ' Q ', ' R ', and that when subjected to a truth-table it yields the following column under its main connective:

P	Q	R	A
T	T	T	T
T	T	F	F
T	F	T	F
T	F	F	T
F	T	T	F
F	T	F	T
F	F	T	F
F	F	F	F

Selecting the assignments for which A comes out true, we may write down corresponding e.c.'s in which a variable appears unnegated if it takes the value T in the assignment and negated if it takes the value F. Thus, corresponding to lines 1, 4, and 6 of the above table, we have

$$' (P \ \& \ Q \ \& \ R) '$$
$$' (P \ \& -Q \ \& -R) '$$
$$' (-P \ \& \ Q \ \& -R) '.$$

199

Forming the disjunction of these three, we obtain a distinguished D.N.F. which, it is fairly evident, is equivalent to the given wff A. Conversely, given a distinguished D.N.F. for a certain wff, each e.c. in it determines an assignment of values to its variables for which the wff is true, and the truth-table for it can be written down. Thus the distinguished D.N.F. of a wff embodies in symbolic form the outcome of a truth-table test on that wff.

Similar remarks can be made about the distinguished C.N.F. Pursuing the above example, corresponding to each assignment for which A comes out false, we may write down e.d.'s in which a variable appears *negated* if it takes the value T in the assignment and *unnegated* if it takes the value F. Thus, corresponding to lines 2, 3, 5, 7, and 8 of the test, we obtain

$$' -P \lor -Q \lor R \,'$$
$$' -P \lor Q \lor -R \,'$$
$$' P \lor -Q \lor -R \,'$$
$$' P \lor Q \lor -R \,'$$
$$' P \lor Q \lor R \,'.$$

Forming the conjunction of these, we obtain a distinguished C.N.F. which is equivalent to the given wff (the reader should test this for himself). Conversely, given a distinguished C.N.F. for a certain wff, each e.d. in it determines an assignment of values to its variables for which the wff is *false* and the truth-table for it can be written down. The distinguished C.N.F., like the distinguished D.N.F., is the symbolic embodiment of a truth-table test.

It is only fair to point out to students that, if they are simply called upon to obtain a normal form equivalent to a given wff, the *quickest* way is usually to perform a truth-table test and read off a normal form from that; the only merit possessed by the burdensome procedures for obtaining normal forms described in this appendix is that they are *independent* of such a test.

The Elementary Theory of Classes

This appendix is intended to give readers a foretaste of what they are likely to find if they pursue logic beyond the confines of this book. Not that anything in it is particularly difficult: indeed, in its first stages the theory of classes is no harder than, and bears a close resemblance to, the propositional calculus, as we shall see. But in its upper stages it raises problems of great interest concerning the foundations of mathematics which, at the date of writing, remain unsolved.

It is not possible to give a precise definition of what a class is. Intuitively, a class is a collection of entities of any kind, and we come to know classes typically in one of two ways: either we are given a *list* of their members, or we are given a *condition* for membership of the class. Let us consider these two ways in turn.

Given a list of things, say Tom, Dick, and Harry, we may consider the class which has just those things as members. Let us agree to name this class by enclosing the names in the list in curly brackets. Thus

'{Tom, Dick, Harry}'

shall be our name for the class whose members are just Tom, Dick, and Harry. Similarly {0,1,2,3,4,5,6,7,8,9} shall be the class whose members are the first ten natural numbers. It is normal to abbreviate ' is a member of ' to the Greek letter ' ε '. Then true propositions will be

Dick ε {Tom, Dick, Harry}

3 ε {0,1,2,3,4,5}

3 + 3 ε {6,7,8,9}.

We more commonly, however, determine classes by stating conditions for membership of them: thus we speak of the class of inhabitants of London, the class of even numbers, the class of female chiropodists, etc. We use this device for describing classes either when it is practically inconvenient to list their members by name (the class of inhabitants of London) or when it is theoretically impossible to do so (the class of even numbers, which is infinite in size). Using the device of variables ' x ', ' y ', ' z ' as in predicate calculus, we may adopt the notation

$$\text{'} \{ x: \ldots x \ldots \} \text{'}$$

as a name for the class of objects x such that $\ldots x \ldots$. Thus $\{x: x$ inhabits London$\}$ will be the class of inhabitants of London and $\{x: x$ is an even number$\}$ will be the class of even numbers. Then true propositions will be

Elizabeth II $\varepsilon \{x: x$ inhabits London$\}$
256 $\varepsilon \{x: x$ is an even number$\}$.

More generally, any property F will determine a class, namely the class of things with property F, or $\{x: Fx\}$. And an arbitrarily selected object a will be a member of this class if and only if it has the property F. In symbols:

(P1) $a \varepsilon \{x: Fx\} \longleftrightarrow Fa.$

(P1) is our first basic principle concerning the membership of classes. It seems hard to doubt its truth but, though we shall accept it for the purposes of this appendix, I shall outline reasons for rejecting it, at least in full generality, at the end.

What constitutes the identity of a class? We want to say, for example, that the class of even numbers is the same as the class of numbers divisible by 2, and that the class $\{2,3,5,7\}$ is the same as the class of prime numbers less than 10. Classes are identical if they have exactly the same members, i.e. if anything which is a member of the one is a member of the other and vice versa. In order to state this principle more strictly, we adopt the Greek letters ' α ', ' β ', ' γ ', . . . as *class-variables*, whose range is understood to be restricted to classes as numerical variables in algebra are understood to have their range restricted to numbers. Then our second principle concerning classes is the following:

(P2) $(x)(x \varepsilon \alpha \longleftrightarrow x \varepsilon \beta) \rightarrow \alpha = \beta.$

If an object is a member of α if and only if it is a member of β, i.e. if α and β have exactly the same members, then by (P2) α and β are identical.

We may illustrate the use of (P1) and (P2) by proving from them the quite trivial result

200 $\vdash \{x : x \varepsilon a\} = a.$

200 merely affirms that the class of things which are members of a class a just *is* a. The proof is easy: as a particular case of (P1), taking 'Fx' as '$x \varepsilon a$', we have

(1) $a \varepsilon \{x : x \varepsilon a\} \longleftrightarrow a \varepsilon a.$

(1) affirms that an object is a member of $\{x : x \varepsilon a\}$ if and only if it is a member of a, i.e. that the two classes have exactly the same members. By (P2), therefore, taking 'a' as $\{x : x \varepsilon a\}$ and 'β' as a, we conclude the identity of the two classes, which is the result 200. Tacitly we are here using a step of UI on (1), followed by a step of MPP in connection with a substitution-instance of (P2). Our proofs in this appendix will in general remain at this informal level.

Certain operations on classes are of great importance: here we introduce the three most basic. First, given two classes a and β, we may define the *union* of a and β as the class of things which are either members of a or members of β. For example, if a is $\{1,2,3\}$ and β is $\{2,3,4\}$, then the union of a and β is $\{1,2,3,4\}$. Or, if a is the class of Englishman and β the class of doctors, then the union of a and β is the class of people who are either Englishmen or doctors (including English doctors). Using a symbol reminiscent of 'v', we write the union of a and β as '$a \cup \beta$'. A formal definition is

$Df. \cup : a \cup \beta = \{x : x \varepsilon a \vee x \varepsilon \beta\}.$

An immediate consequence of the definition by (P1) (taking 'Fx' as '$x \varepsilon a \vee x \varepsilon \beta$') is

201 $\vdash a \varepsilon a \beta \longleftrightarrow a \varepsilon a \vee a \varepsilon \beta.$

Concerning the union of classes, the following results hold:

202 $\vdash a \cup \beta = \beta \cup a;$

203 $\vdash (a \cup \beta) \cup \gamma = a \cup (\beta \cup \gamma);$

204 $\vdash a \cup a = a.$

Of these, I prove only 202; the proofs of 203 and 204 are entirely similar. By 201,

(2) $a \varepsilon a \cup \beta \longleftrightarrow a \varepsilon a \vee a \varepsilon \beta$

(3) $\qquad\qquad \longleftrightarrow a \varepsilon \beta \vee a \varepsilon a$

(4) $\qquad\qquad \longleftrightarrow a \varepsilon \beta \cup a.$

Line (3) follows from line (2) by elementary propositional calculus reasoning, and line (4) from line (3) by an application of 201 taking 'α' as β and 'β' as α. The device of 'stacking' biconditionals should be self-explanatory. Since, by (4), $\alpha \cup \beta$ and $\beta \cup \alpha$ have exactly the same members, they are identical by (P2).

Second, given two classes α and β, we define the *intersection* of α and β as the class of things which are both members of α and members of β. For example, if α is $\{1,2,3\}$ and β is $\{2,3,4\}$ then the intersection of α and β is $\{2,3\}$. Or, if α is the class of Englishmen and β the class of doctors, then the intersection of α and β is the class of male English doctors. We write the intersection of α and β as '$\alpha \cap \beta$'. Thus:

$$Df. \cap: \quad \alpha \cap \beta = \{x: x \, \varepsilon \, \alpha \, \& \, x \, \varepsilon \, \beta\}.$$

As an immediate consequence by (P1), we have

205 $\vdash a \, \varepsilon \, \alpha \cap \beta \longleftrightarrow a \, \varepsilon \, \alpha \, \& \, a \, \varepsilon \, \beta$,

and the analogues of 202–204 can also very easily be proved:

206 $\vdash \alpha \cap \beta = \beta \cap \alpha$;

207 $\vdash (\alpha \cap \beta) \cap \gamma = \alpha \cap (\beta \cap \gamma)$;

208 $\vdash \alpha \cap \alpha = \alpha$.

In view of 202–204 and 206–208, union and intersection of classes, in the terminology of the previous appendix, are commutative, associative, and idempotent.

Third, for any class α we define its *complement* as the class of things which are not members of α. Of course, in any interpretation of the theory of classes, as in the case of the predicate calculus, we have a *fixed universe of discourse* in mind, and the complement of a class is relative to that universe. Thus, given the universe of positive integers, the complement of the class of even numbers is the class of odd numbers (and vice versa), and, given the universe of human beings, the complement of the class of males is the class of females (and vice versa). Given the universe of natural numbers, the complement of the class $\{0,1,2\}$ is the class of numbers greater than 2. We denote the complement of α by '$\alpha^{\mid}$'. Thus:

$$Df. ^{\mid}: \quad \alpha^{\mid} = \{x: -(x \, \varepsilon \, \alpha)\}.$$

An immediate consequence of this definition by (P1) is

209 $\vdash a \, \varepsilon \, \alpha^{\mid} \longleftrightarrow -(a \, \varepsilon \, \alpha)$.

A law of double negation for class is forthcoming in the form

210 $\vdash a^{||} = a.$

Union, intersection, and complementation of classes are related to one another by various analogues of de Morgan's laws. Here we state and prove only one; the reader should be able to compose and prove the remainder for himself, if he wishes.

211 $\vdash (a \cap \beta)^{|} = a^{|} \cup \beta^{|}.$

By 209,

$$(5) \quad a \, \varepsilon \, (a \cap \beta)^{|} \longleftrightarrow -(a \, \varepsilon \, a \cap \beta)$$

$$(6) \qquad\qquad\qquad \longleftrightarrow -(a \, \varepsilon \, a \, \& \, a \, \varepsilon \, \beta)$$

$$(7) \qquad\qquad\qquad \longleftrightarrow -(a \, \varepsilon \, a) \, \mathrm{v} \, -(a \, \varepsilon \, \beta)$$

$$(8) \qquad\qquad\qquad \longleftrightarrow a \, \varepsilon \, a^{|} \, \mathrm{v} \, a \, \varepsilon \, \beta^{|}$$

$$(9) \qquad\qquad\qquad \longleftrightarrow a \, \varepsilon \, a^{|} \cup \beta^{|}.$$

Here (6) follows from (5) by 205, (7) from (6) by propositional calculus reasoning, (8) from (7) by 209 again, and (9) from (8) by 201. 211 now follows by (P2), since we have shown that $(a \cap \beta)^{|}$ and $a^{|} \cup \beta^{|}$ have exactly the same members.

So far, we have used curly brackets round lists of names *informally* as designations for classes; but it is possible to introduce this device *formally* by defining it in terms of our other use of curly brackets, where a condition for membership of a class is given. For example, {Tom, Dick} may be defined as the class whose members *are either Tom or Dick*. In general, we may put

$$\{a, b\} = \{x : x = a \, \mathrm{v} \, x = b\}.$$

Then $\{a, b\}$ will be the class whose members are just a and b, i.e. the class of things which either are a or are b. To be a member of $\{a, b\}$ is to be either a or b:

212 $\vdash c \, \varepsilon \, \{a, b\} \longleftrightarrow c = a \, \mathrm{v} \, c = b.$

We may call $\{a, b\}$ the *pair class* of a and b. Since both a and b satisfy the right-hand side of 212, we have obviously

213 $\vdash a \, \varepsilon \, \{a, b\} \, \& \, b \, \varepsilon \, \{a, b\}.$

This device can obviously be extended to classes with three or more members. It can also be extended in the other direction to classes with exactly *one* member. Thus $\{a\}$ will be the class whose sole member is a. We may define

$$\{a\} = \{x: x = a\},$$

and then it will follow that

214 $\vdash b \, \varepsilon \, \{a\} \longleftrightarrow b = a$

215 $\vdash a \, \varepsilon \, \{a\}$.

$\{a\}$ may be called the *unit class* of a. The unit class of a is *not* the same as a itself; we shall see later the importance of distinguishing the two.

It may happen that, when we state a condition for membership of a class, it turns out that *nothing* satisfies that condition. What are we to say about the class in this case? What, for example, about $\{x: x \text{ is a unicorn}\}$, the class of unicorns; or, in general, what about $\{x: Fx\}$ when $(x) - Fx$? It proves theoretically simplest to allow in this case that the *class* exists and to call it *empty*. There are no unicorns; but there is a class of unicorns, and precisely one way of saying that there are no unicorns is to say that this class is empty.

It is a (perhaps surprising) consequence of (P2) that any two empty classes are *identical*. For two classes are identical if they have exactly the same members, i.e. if there is no member of one which is not a member of the other. But if two classes are empty, then there is no member of one not a member of the other, just because there is no member of either. Therefore they are identical. Put a little more formally:

216 $(x) - (x \, \varepsilon \, a), (x) - (x \, \varepsilon \, \beta) \vdash (x)(x \, \varepsilon \, a \longleftrightarrow x \, \varepsilon \, \beta)$.

This sequent is provable very simply using predicate calculus rules, and is clearly related to 133, one of the paradoxes of formal implication. Hence by (P2)

217 $(x) - (x \, \varepsilon \, a), (x) - (x \, \varepsilon \, \beta) \vdash a = \beta$.

In virtue of 217, we may speak of *the* empty class, or, as it is more commonly called, the *null class*. We shall denote it by ' Λ ', and we may define it by stating any condition which we know nothing satisfies, say that of being not identical with oneself. Thus

$$\Lambda = \{x: x \neq x\}.$$

The null class is the class of x such that x is not identical with itself. Since, as a matter of logic, $(x)(x = x)$, it is easy to prove by (P1)

218 $\vdash (x) - (x \,\varepsilon\, \Lambda)$.

The intersection of a class with its own complement is empty, as we should expect. Thus we can prove

219 $\vdash a \cap a^{|} = \Lambda$.

For suppose $a \,\varepsilon\, a \cap a^{|}$. Then by 205 $a \,\varepsilon\, a \,\&\, a \,\varepsilon\, a^{|}$, whence by 209 $a \,\varepsilon\, a \,\&\, -(a \,\varepsilon\, a)$, which is a contradiction. Thus $(x)-(x \,\varepsilon\, a \cap a^{|})$, whence 219 follows from 217 and 218.

The complement of the null class will clearly be the class of which *everything* is a member; thus, for any given interpretation in some universe of discourse, it will be the class of things in that universe. We may call it the *universe class*, denote it by ' V ', and define it

$$V = \{ x : x = x \}.$$

Thus V is the class of things identical with themselves, and, since as a matter of logic $(x)(x = x)$, it is easy to prove by (P1)

220 $\vdash (x)(x \,\varepsilon\, V)$.

It should be obvious that, in virtue of (P2), any two classes which have everything as members are identical. We have

221 $(x)(x \,\varepsilon\, a), (x)(x \,\varepsilon\, \beta) \vdash a = \beta$.

Using 220 and 221, we may show that the union of a class with its own complement is, as we should expect, the universe class:

222 $\vdash a \cup a^{|} = V$.

An important relation that may exist between two classes is that of *inclusion*. A class a is *included* in a class β if all the members of a are members of β. We then write ' $a \subseteq \beta$ ', and adopt the definition

$$Df. \subseteq : \quad a \subseteq \beta \longleftrightarrow (x)(x \,\varepsilon\, a \rightarrow x \,\varepsilon\, \beta).$$

The inclusion-relation must be carefully distinguished from the membership-relation. Suppose I am an inhabitant of Chelsea. Then I am a *member* of the class of inhabitants of Chelsea. Further, since all inhabitants of Chelsea are inhabitants of London, the class of inhabitants of Chelsea is *included* in the class of inhabitants of

London, so that I am a member of this latter class too. But the class of inhabitants of Chelsea is not a *member* of the class of inhabitants of London; if it were, it would be an inhabitant of London; but classes, unlike people, are not inhabitants of any place. Again, the class of whales is included in the class of mammals, but it is not a member of that class, since, being a class, it is not a whale. If Rufus, however, is a whale, then it is a member of the class of whales and so a member of the class of mammals. But Rufus, not being a class, is not included in the class of mammals, any more than I, if I live in Chelsea, am included in the class of inhabitants of London, though I am a member of that class.

Hence arises the importance of distinguishing objects from their unit classes. For if I am a member of the class of inhabitants of Chelsea, then my unit class is *included* in that class, since all its members (in this case only one) are members of the class. We have in fact

223 $\vdash a \, \varepsilon \, a \longleftrightarrow \{a\} \subseteq a.$

An object is a member of a class if and only if its unit class is included in that class. Inclusion is a relation between *classes*: membership is (typically) a relation between an individual and a class.[1]

The more obvious properties of the inclusion relation are given in the following theorems, which in general follow immediately from its definition.

224 $\vdash a \subseteq a;$

225 $\vdash a \subseteq \beta \,\&\, \beta \subseteq \gamma \rightarrow a \subseteq \gamma;$

226 $\vdash a \subseteq \beta \rightarrow \beta^{\mathsf{I}} \subseteq a^{\mathsf{I}};$

227 $\vdash \Lambda \subseteq a;$

228 $\vdash a \subseteq V.$

224–226 are, in a sense, theorems of predicate calculus wearing a thin disguise. For example, when unabbreviated, 224 is merely the law of identity $(x)(x \, \varepsilon \, a \rightarrow x \, \varepsilon \, a)$. Similarly, 227 and 228 (the null class is included in any class, and any class is included in the universe

---

[1] Of course, we do not rule out the possibility of classes with classes as *members*: indeed, the advanced theory of classes would have little interest if such classes were not admitted.

class) are class-analogues of the paradoxes of formal implication 133 and 132. Slightly harder to prove are the following:

229 $\vdash a \subseteq \beta \& \beta \subseteq a \longleftrightarrow a = \beta$;

230 $\vdash a \subseteq \beta \longleftrightarrow a \cup \beta = \beta$;

231 $\vdash a \subseteq \beta \longleftrightarrow a \cap \beta = a$.

I give an informal proof of 231. First suppose $a \subseteq \beta$, and let $a \, \varepsilon \, a$. Then clearly $a \, \varepsilon \, \beta$, and so $a \, \varepsilon \, a \cap \beta$. Conversely, if $a \, \varepsilon \, a \cap \beta$, then $a \, \varepsilon \, a$ anyway. So that $a \, \varepsilon \, a \cap \beta$ if and only if $a \, \varepsilon \, a$, whence by (P2) $a \cap \beta = a$. This proves the conditional from left to right. Now suppose $a \cap \beta = a$, and let $a \, \varepsilon \, a$. Then $a \, \varepsilon \, a \cap \beta$, whence $a \, \varepsilon \, \beta$. Hence any member of a is a member of β, so that $a \subseteq \beta$. This proves the conditional from right to left, and completes the proof of 231.

I shall not here develop further the elementary theory of classes. But the reader will no doubt have detected many analogies between this theory and the propositional calculus (compare 231, for example, with the tautology '$P \rightarrow Q \longleftrightarrow (P \& Q \longleftrightarrow P)$'), and he may, if he wishes, devise and prove other theorems for classes on the basis of this comparison; or he may profitably consult Suppes [25]. I would rather conclude on a note of hesitancy. For, though it may appear that our theory of classes has proceeded smoothly enough from intuitively acceptable assumptions, there are certain difficulties that must be overcome if we are to have a consistent and workable system.

Consider, first, the apparently inoffensive result 220, $(x)(x \, \varepsilon \, V)$. Are we to allow, as a consequence by UE, $V \, \varepsilon \, V$? If the universe class is the class of which *everything* is a member, as we defined it to be, then is it a member of itself? Perhaps we might allow this as a theorem[1]; but there is certainly something intuitively queer about a class which has itself as a member. Certainly, all classes we ordinarily consider are *not* self-membered: the class of whales, for example, is not itself a whale, the class of natural numbers is not itself a natural number. The question as to whether there are self-membered classes leads us to consider the *class* of classes which are not self-membered, or, in our curly-brackets notation

$$\{x: -(x \, \varepsilon \, x)\}$$

[1] It is a theorem of the theory of classes in Quine [18]; but it is rejected in most theories of classes.

—the class of objects x such that x is not a member of itself. Let us call this class R. Then, as we have seen, most classes we normally handle are members of R.

In virtue of (P1), the condition for membership of R is given by the biconditional

(10) $a \, \varepsilon \, \{x: -(x \, \varepsilon \, x)\} \longleftrightarrow -(a \, \varepsilon \, a)$,

or

(11) $a \, \varepsilon \, R \longleftrightarrow -(a \, \varepsilon \, a)$.

But (11) holds for arbitrary a: taking a as R itself, we have

(12) $R \, \varepsilon \, R \longleftrightarrow -(R \, \varepsilon \, R)$.

(12) leads directly to a contradiction. This contradiction was first discovered by Russell, and the paradox obtained by considering the class of classes which are not members of themselves is known as Russell's paradox.

It is important to stress that the implicit contradiction (12) has been obtained by entirely elementary reasoning from (P1). And it is no exaggeration to say that modern theories of classes all have as their starting-point some device for the avoidance of paradoxes such as Russell's. Russell's own escape-route was, essentially, to rule out expressions such as '$a \, \varepsilon \, a$' as not well-formed and to develop a hierarchy of classes in such a way that the membership-relation can only meaningfully be said to hold between an object at one level and a class at the next level up. The resulting *theory of types* has a certain naturalness, but is awkward and clumsy to work with in practice. Other theories accept '$a \, \varepsilon \, a$' as well-formed (though in general false), and avoid contradictions by imposing further conditions on the right-hand side of (P1). Yet other theories seek to avoid the Russell paradox by questioning the assumption behind (P1) that *every* predicate 'F' determines a class $\{x: \neq Fx\}$. There is as yet no general agreement amongst logicians as to the best or the proper way of avoiding Russell's paradox. But one thing at least is clear: (P1), despite its intuitive attractiveness and apparent self-evidence, cannot be accepted as it stands. And it is one merit of modern symbolic logic to have shown just this. There may also be a philosophical moral to be drawn concerning the dangers of over-reliance on intuitions; but this is not the place to draw it.

BIBLIOGRAPHY

The following bibliography is highly selective. It is intended primarily as a starting-point for the student who wishes to go deeper into logic and is uncertain where to begin. Thus it should be used in conjunction with the notes following it, which indicate some of the possible directions in which the reader might now move. Only books are mentioned here; none the less the vast majority of recent logical work has proceeded by means of articles in journals. The student aiming to encompass these should consult the *Journal of Symbolic Logic*, which began in 1936 with a complete bibliography of earlier work and has kept up to date since then by means of periodic bibliographical supplements.

[1] BASSON, A. H., and O'CONNOR, D. J., *Introduction to Logic* (2nd ed.), London, 1957.

[2] CHURCH, A., *Introduction to Mathematical Logic*, I, Princeton, 1956.

[3] COPI, I. M., *Symbolic Logic*, New York, 1954.

[4] FITCH, F. B., *Symbolic Logic*, New York, 1951.

[5] FREGE, G., *The Foundations of Arithmetic* (translated by J. L. Austin), Oxford, 1950.

[6] FREGE, G., *Translations From the Philosophical Writings of Gottlob Frege* by Peter Geach and Max Black, Oxford, 1952.

[7] HILBERT, D., and ACKERMANN, W., *Principles of Mathematical Logic*, New York, 1950.

[8] JOSEPH, H. W. B., *An Introduction to Logic*, Oxford, 1906.

[9] KALISH, D., and MONTAGUE, R., *Logic: Techniques of Formal Reasoning*, New York, 1964.

[10] KLEENE, S. C., *Introduction to Metamathematics*, Princeton, 1952.

[11] KNEALE, WILLIAM and MARTHA, *The Development of Logic*, Oxford, 1962.

[12] ŁUKASIEWICZ, J., *Aristotle's Syllogistic* (2nd ed.), Oxford 1957.

[13] MATES, B., *Stoic Logic*, Berkeley and Los Angeles, 1961.

[14] MATES, B., *Elementary Logic*, New York, 1965.

[15] MENDELSON, E., *Introduction to Mathematical Logic*, Princeton, 1964.

[16] PRIOR, A. N., *Formal Logic* (2nd ed.), Oxford, 1962.

[17] QUINE, W. V., *Methods of Logic*, New York, 1950.

[18] QUINE, W. V., *Mathematical Logic* (revised ed.), Cambridge, Mass., 1951.

Bibliography

[19] QUINE, W. V., *Set Theory and Its Logic*, Cambridge, Mass., 1963.

[20] ROSENBLOOM, P., *Elements of Mathematical Logic*, New York, 1950.

[21] RUSSELL, B., *Introduction to Mathematical Philosophy*, London, 1919.

[22] STEBBING, L. S., *A Modern Introduction to Logic*, London, 1930.

[23] STRAWSON, P. F., *Introduction to Logical Theory*, London, 1952.

[24] SUPPES, P., *Introduction to Logic*, Princeton, 1957.

[25] SUPPES, P., *Axiomatic Set Theory*, Princeton, 1960.

[26] TARSKI, A., *Logic, Semantics, Metamathematics* (papers from 1923 to 1938, translated by J. H. Woodger), Oxford, 1956.

[27] WHITEHEAD, A. N., and RUSSELL, B., *Principia Mathematica*, vols. I–III, Cambridge, 1910–13; abridged text of vol. I, Cambridge, 1962.

Notes

1. Students wishing to pursue the topics and methods of this book in greater detail should try [3], [4], [9], [14], [17], [24]. [24] will seem most familiar, and goes a good deal further than my treatment in many respects; [4] is closest in terms of the actual rules of derivation employed. [3], [9], and [24] are rich in exercises which can be used as a supplement to this course. [17] has a nice treatment of truth-table testing, but I personally don't like its quantifier rules. [9] is up to date and very accurate, but a bit hard. [14] is a very thorough and lucid survey of modern techniques in logic.

2. Students interested in the propositional calculus from standpoints other than the present one (e.g. an axiomatic approach rather than a natural-deduction one) should consult [1], [7], [16], or, best of all, [2], which is a *very* comprehensive treatment, though hard going.

3. Students interested in the predicate calculus from other standpoints might try [7], [14], [15], [18]. Again, a comprehensive treatment is in [2].

4. For traditional logic, use [8] if you want to be thorough, [22] if you want to be brisk, and [12] if you want to be up to date and (largely) accurate. [13] is an excellent account of the logic of the Stoics, who anticipated many aspects of the propositional calculus as well as the philosophical theories of Frege.

5. The history of logic is probably best studied in [11]; [12] and [13] form useful supplements on ancient logic, and there are interesting historical comments throughout [16]. [5] and [27] (especially the introductory chapters) are modern classics.

6. Those interested in philosophical issues related to formal logic should read [5], [6], [21], and [23]. The first chapter (Chapter 0) of [2] deserves to be read several times by *all* philosophers.

7. For the theory of classes, begin with [25]; then use [18] and [19]. These will suggest further lines of study.

8. For other aspects of more advanced logic, [10] is indispensible, but difficult for those without a mathematical background. [15] is also useful —more up to date in some ways than [10], but less accurate and *very* condensed ([10] also contains a good chapter on the theory of classes).

212

[20] goes a long way in a few pages, but will appeal more to mathematicians than philosophers. [26] gives an excellent idea of the kind of work that went on in logic between the wars, and many of the papers in it are foundation-stones for much contemporary work. For some interesting by-ways in modern logic (e.g. many-valued logics, modal logics), begin with [16], which has a useful bibliography for further reading.

9. If you are in this thing seriously, you will need, more than anything else, [2] and [10]. If you find mathematical proofs difficult, try a change of pace—read them slowly and several times over, at first fairly casually, later not leaving a sentence until you are *certain* you understand it. If it helps, bear in mind that *professional mathematicians* find them hard too. A good proof can be savoured in much the way a professional game of chess is by enthusiasts, and in a similar manner records a fragment of intellectual history: don't rush it.

10. Logic texts display a rich variety of different symbols for the same operations, and there is little sign of any growing uniformity. This is something one must learn to live with.

List of Logical Symbols and Abbreviations with Main Text References

INDEX

A : *see* rule of assumptions
abstract noun(s), 160
Ackermann, W., 157*n*.
affirming the consequent, fallacy of, 17–18
A-form proposition(s), 169ff., 175, 177
algebra, 49, 56, 71, 94, 98, 104, 156, 202
algebraic expression(s), 64
' all ', 92–102, 96
ancient logic, 212
' and ', 6, 19 ; *see also* conjunction, ' & '
antecedent, 7
    fallacy of denying, 17–18
antisymmetric, 181, 182, 188
' any ', 96, 101
arbitrarily selected object(s), 107, 111, 112, 116
arbitrary name(s), ix, 107, 114, 115, 129–30, 138, 139, 155
argument(s), 1ff., 5, 6, 12, 167
    circular, 34
    pattern(s) of, *see* pattern(s) of argument
    and proposition(s), 52
    soundness, 1–2, 5, 8, 81 ; *see also* soundness
argument-frame, 12, 52
Aristotelian theory of the syllogism, 94, 169, 171 ; *see also* theory of the syllogism
Aristotle, 94, 169, 173, 174
arithmetic, 39, 192
assertion-sign, 11, 48, 50
associative laws, 191ff.
assumption(s), 8–9, 12, 52
    discharged, 15
    existential, *see* existential
    and premiss(es), 8
    recorded in proof, viii, 8–9, 10
    rule of, *see* rule of
    special, 109
asymmetric, 180–1, 182, 183, 184, 185, 186, 188
' at least *n* ', 165
' at least one ', 97, 165
' at least two ', 97, 164
' at most one ', 165
atom, 189

atomic sentence, 139–40
    and identity, 161
axiomatic development of a calculus, viii, 212

Basson, A. H., 189*n*.
Bernays, P., ix
biconditional, 29–33 ; *see also* ' if and only if ', ' $\longleftrightarrow$ '
bracket(s), 7, 43, 46–7, 139
    curly, *see* curly

calculus, intuitionist, ix
    minimal, ix
    predicate, *see* predicate
    propositional, *see* propositional
    sentential, 42 ; *see also* propositional calculus
canonical
    conjunctive normal form, 198ff.
    disjunctive normal form, 198ff.
C.C.N.F., *see* canonical
C.D.N.F., *see* canonical
Church, A., ix, 152*n*., 157*n*., 158
circular argument, 34
class(es), 201
    complement, *see* complement
    determination of, 201ff.
    empty, 206 ; *see also* null
    and identity, 202
    inclusion, *see* inclusion-relation
    intersection, *see* intersection
    membership, *see* membership-relation
    null, 206–7 ; *see also* empty class
    operation(s) on, 203ff.
    pair, 205
    and property, 202
    and propositional calculus, *see* theory of classes
    self-membered, 209–10
    theory of, *see* theory of
    union, *see* union
    unit, *see* unit class
    universe, 207
class-variable(s), 202

217

Index

' is a member of ', 201
' is implied by ', 70
' is the same object as ', 160
' it ' (and individual variables), 95
integer, non-negative, 105
  positive, 105
interderivability, 34–5
internal structure (of a proposition), 93, 167
interpretation, 156
  true under in non-empty universe(s), 156–7
intersection, 204, 205
intransitive, 183, 185, 188
introduction, rules of, 19
  existential quantifier introduction, *see* rule of
  identity introduction, *see* rule of
  sequent introduction, *see* rule of
  theorem introduction, *see* rule of
  universal quantifier introduction, *see* rule of
  &-introduction, *see* rule of
  v-introduction, *see* rule of
intuitionist propositional calculus, ix
  predicate calculus, ix
invalid, *see* valid, validity
invalidity, 64, 81 ; *see also* validity

Johannson, I., ix*n*.
Joseph, H. W. B., 169*n*., 171*n*.

language, formal, 42ff.
law(s), associative, 191ff.
  commutative, 191ff.
  conversion, 170–1, 177
    per accidens, 171
    simple, 171
  de Morgan's, *see* de Morgan's
  distributive, *see* distributive
  of double negation, 52, 191, 205
  of excluded middle, ix, 52–3, 64–5, 91, 149
  of idempotence, 191ff.
  of identity, 52, 149, 208
  logical, *see* logical law(s)
  of non-contradiction, 52, 149
  of obversion, 171
lemma, 84
letter, 150
  predicate, *see* predicate-letter(s)
  schematic, *see* schematic
logic(s), ancient, 212
  defined, 5
  history of, 212
  many-valued, 213
  and mathematics, *see* mathematics
  modal, 213

nature of, 1–5
Stoic, 212
see also contemporary, mathematical, symbolic, traditional logic
logical
  connective(s), 43, 65, 69, 139
    main, *see* main connective
    ranked, 46–7
    scope, *see* scope
    subordination, *see* subordination
  form, 4, 5, 7–8, 55–6, 69, 92, 93, 167 ; *see also* pattern(s) of argument
  law(s), 52, 69
  notation, 3, 4, 39, 167 ; *see also* logical symbol(s), symbolism
  relation(s), 69–71, 169
  symbol(s), 3, 4 ; *see also* logical notation, symbolism
  system, 80, 90, 186
  truth(s), 52, 69, 149
logically true, 52, 69
  and interpretation, 157
  valid, 157
Łukasiewicz, J., 174*n*.

main column, 66
  connective, 48, 66
major premiss, 171, 179
  term, 171
many-valued logics, 213
material implication, 60–1, 154
  paradoxes of, *see* paradox(es)
Mates, B., viii*n*., ix*n*.
mathematics, 77, 91
  foundations of, 201
  and logic, 3, 42, 186
mathematical concepts, 186
  functions, 71
  induction, 85 ; *see also* proof by induction
  logic, 3 ; *see also* contemporary, symbolic logic
  symbol, 4
matrix(es), 65, 156
mechanical check of proof, 39, 67, 158
  proof-discovery, 90, 91, 158
  test, 157
member, 207–8
membership-relation, 186, 202, 207–8
metatheorem, 77
Metatheorem I, 77, 81, 85 ; Metatheorem II, 83–4, 88, 91 ; Metatheorem III, 84, 89, 91
metalogical variable, 49, 139
minimal calculus, ix
minor premiss, 171
  term, 171
modal logics, 213
modi, 61–2

Index

proof—*contd*
mechanical check of, discovery of, *see* mechanical
procedure, viii, 8–9, 10
stages of, 8, 77
proof-discovery, vii, 62, 67, 90, 91, 114–15, 158
proper name(s), 93, 114, 138, 139, 160, 166, 167, 179
property, 98, 149, 179
and class, 202
property-expression(s), 93
proposition(s) (or statement(s)), 1, 69–71, 93, 141
A-, E-, I-, O- form, *see* A-, E-, I-, O- form
and argument, 52
conditional, *see* conditional
existential, *see* existential
internal structure, *see* internal
particular affirmative, negative, 169
and propositional function(s), 143, 179
and sentence, 6, 55–6, 141, 179
and truth, 2
universal, *see* universal
affirmative, negative, 169
propositional calculus, 42, 93, 94, 104, 139, 167, 212
complete, *see* completeness
consistent, *see* consistency
intuitionist, ix
and predicate calculus, 93, 94, 104, 139, 140, 145, 157, 158
rules, *see* rule(s)
and theory of classes, *see* theory of connective(s), *see* logical connective(s)
function, 141, 143–4
and properties and relations, 149ff., 179
and propositions, 143, 179
variable(s), 7, 43, 49, 55–6, 139, 140, 156, 189
proved sequent(s), 12 ; *see also* derivable, valid sequent(s)

quality, 178
quantifier(s), 123, 143, 179
existential, *see* existential
notation, translation into, *see* translation
numerically definite, 165–6
rules, ix, 104, 145–6
scope of, 143–4
universal, *see* universal
vacuous, *see* vacuous
quantifier-shift principle(s), 128–30

quantity, 178
Quine, W. V. O., 158*n*., 163*n*., 166*n*., 209*n*.

(R), 193
RAA, *see* rule of *reductio ad absurdum*
range, 187
rank, of connectives, 46–7
rational numbers, 186
real numbers, 186
reductio ad absurdum, rule of ; *see* rule of
reduction to the first figure, 174
reflexive, 183, 184, 187, 188
relation(s), 98, 128, 149, 159, 177, 179, 179–88, 186
antisymmetric, *see* antisymmetric
asymmetric, *see* asymmetric
dyadic, 180
inclusion-, 207–8
intransitive, *see* intransitive
irreflexive, *see* irreflexive
logical, *see* logical relation(s)
membership-, *see* membership-relation
non-reflexive, *see* non-reflexive
non-symmetric, *see* non-symmetric
non-transitive, 183
properties of, 177, 179–88
reflexive, *see* reflexive
serial, 187
symmetrical, *see* symmetric
transitive, *see* transitive
relational expression(s), 98, 179
sentence(s), 98
relationship(s), logical, *see* logical relation(s)
relative clauses, 101
rest on, 8, 9
reverse-E, 139
replacement, principle of ((R)), 193
rule(s)
complete (propositional calculus), 42, 83 ; *see also* completeness
of derivation (inference), 3, 6, 8, 39, 64, 69, 91
propositional calculus, viii ; résumé of, 39–40
predicate calculus, ix, 145
with identity, *cf.* 161
recorded in proof, 8, 10
derived, *see* derived
formation, *see* formation
primitive, *see* primitive
of quality, 178
quantifier, *see* quantifier
of quantity, 178
safe, 42, 75ff., 81, 90 ; *see also* consistency

222